One Hundred Years of Pressure

Archimedes

NEW STUDIES IN THE HISTORY AND PHILOSOPHY OF SCIENCE AND TECHNOLOGY

VOLUME 51

Archimedes has three fundamental goals; to further the integration of the histories of science and technology with one another: to investigate the technical, social and practical histories of specific developments in science and technology; and finally, where possible and desirable, to bring the histories of science and technology into closer contact with the philosophy of science. To these ends, each volume will have its own theme and title and will be planned by one or more members of the Advisory Board in consultation with the editor. Although the volumes have specific themes, the series itself will not be limited to one or even to a few particular areas. Its subjects include any of the sciences, ranging from biology through physics, all aspects of technology, broadly construed, as well as historically-engaged philosophy of science or technology. Taken as a whole, *Archimedes* will be of interest to historians, philosophers, and scientists, as well as to those in business and industry who seek to understand how science and industry have come to be so strongly linked.

More information about this series at http://www.springer.com/series/5644

Alan F. Chalmers

One Hundred Years of Pressure

Hydrostatics from Stevin to Newton

Springer

Alan F. Chalmers
Unit for History and Philosophy of Science
University of Sydney
Sydney, NSW, Australia

ISSN 1385-0180 ISSN 2215-0064 (electronic)
Archimedes
ISBN 978-3-319-56528-6 ISBN 978-3-319-56529-3 (eBook)
DOI 10.1007/978-3-319-56529-3

Library of Congress Control Number: 2017936360

Printed on acid-free paper

This Springer imprint is published by Springer Nature
The registered company is Springer International Publishing AG
The registered company address is: Gewerbestrasse 11, 6330 Cham, Switzerland

Acknowledgements

I have been fortunate to have had the flourishing Unit for History and Philosophy of Science at the University of Sydney as a convivial and inspirational home base for the conduct of my research. Three members of that unit deserve special mention. John Schuster gave me encouragement and very detailed commentaries on early versions of my research. It was he that first raised the possibility that my findings might usefully be made available in the form of a short book. For her part, Debbie Castle has given me much logistical support and has been ever ready and willing to compensate for my rudimentary word-processing and IT skills. Jenny Tomlinson drew on her professional expertise to fine-tune the final version of my manuscript, suggesting stylistic improvements and indicating ways in which I might avoid unnecessary repetition. The Unit for HPS gave me an academic base from which to apply for a research grant from the Australian Research Council. My work in the years 2012–2015 was supported by Discovery Grant #DP110102471.

I was privileged to be able to conduct some of my research at the Max Planck Institute for the History of Science in Berlin on three separate occasions from 2014 to 2016 for a total of 8 months. I received crucial encouragement and support from Jürgen Renn. A number of scholars at the Max Planck were much attuned to the work I was doing, and what I learnt from them proved to be invaluable. The incumbents refer to what they do at that institute as 'historical epistemology'. Somewhat like Monsieur Jourdain in Molière's *The Bourgeois Gentleman*, who was very self-satisfied to learn that he had been writing prose for 40 years, it was an eye-opener for me to realize that I have been practising historical epistemology for a similar duration. On the occasion of my latest visit, I turned up with a draft of this book. I had the luxury of being able to discuss the chapters at weekly meetings of a small but dedicated group of knowledgeable scholars, reworking my chapters as we went along. Three months later, exhausted and elated, I had a revised version ready to be sent to the publishers. Of the scholars who gave me help, Christoph Lehner and Pietro Omodeo supplied extensive commentaries on all of my chapters, and Ursula Klein and Wolfgang Lefèvre were also generous with their time and helpful with their comments. Lindy Divarci was on hand to help me with approaches to publishers. I was indeed fortunate to receive this degree of support.

I was a guest in the Philosophy Department at the University of Durham on a number of occasions and participated in workshops connected with my research. I received particularly helpful feedback from Robin Hendry, Peter Vickers and David Wootton. I have also read papers at the University of Bristol, benefitting from critical comments and encouragement from Andrew Pyle and James Ladyman. Both Keith Hutchison and Alan Shapiro gave me detailed comments on complete drafts of my book, the criticisms in each case being especially helpful and constructive because of their bluntness. I gratefully acknowledge the help I have received with translations from Latin from John Schuster, Pietro Omodeo, Christoph Lehner and Michael Hanaghan, French from Pierrick Bourrat and Dutch from Hans Pols and Joyce Van Leeuven. I thank Peter Saw for preparing Figs. 1.1, 6.2 and 6.3 and Georg Pflanz for preparing Figs. 4.1 and 5.2. Others who have given assistance and encouragement warranting acknowledgement are Peter Anstey, Jed Buchwald, Stephen Gaugroger, Antoni Malet, Peter McLaughlin and Matthias Schemmel.

I acknowledge permission to include in this book versions of material published in two journals. The articles in question are 'Intermediate causes and explanations: The key to understanding the Scientific Revolution', *Studies in History and Philosophy of Science*, 43 (2012), 551–562; 'Qualitative novelty in seventeenth-century science: Hydrostatics from Stevin to Pascal', *Studies in History and Philosophy of Science*, 51 (2015), 1–10; and 'Robert Boyle's mechanical account of hydrostatics and pneumatics: Fluidity, the spring of the air and their relationship to the concept of pressure', *Archive for History of Exact Sciences*, 69 (2015), 429–454.

Heinz Post still exerts a strong influence on my research, notwithstanding the fact that he has been dead for some time. He trained me in history and philosophy of science and supervised my PhD thesis. I feel him looking over my shoulder as I write, asking 'what is the point of this study?', 'what does that mean?', 'is that relevant?' and the like. A second important person who is far from dead and whose influence can be measured by the degree of discomfort to which it leads is my partner Sandra Grimes. Her goading of me to continue at difficult junctures in my work has given me someone to blame for the stress involved.

Sydney, NSW, Australia
December, 2016

Alan F. Chalmers

Contents

Chapter 1
Liquids: A Challenge for Seventeenth-Century Mechanics

Abstract The construction of a science of hydrostatics in the seventeenth century came about by way of the replacement of a common sense of pressure by a technical sense of pressure. This proved to be a move that was far from obvious and much can be learnt from it about the significant changes in science that collectively have become known as the Scientific Revolution. The new hydrostatics was defended by appeal to observation and experiment rather than by deriving it from unproblematic postulates granted at the outset, and, in that respect, differed from the science of simple machines that had preceded it. The construction of the technical concept of pressure was a theoretical move, but took place at the level of what could be observed and manipulated experimentally rather than at the level of corpuscles having shape, size and motion that mechanical philosophers of the time came to regard as the ultimate components of the material world.

1.1 Introductory Remarks

Making a distinction between liquids and solids is as old as language itself.[1] However, the challenge of acquiring a theoretical grasp of that distinction adequate for a science of hydrostatics proved to be a considerable one. This book contains an historical account and an analysis of how that challenge was met during the course of the seventeenth century.

The extension of the mechanics of simple machines to include machines involving bodies of liquid at rest was far from straightforward. The task proved to require the development of a technical, as distinct from a common sense, notion of pressure. The degree of conceptual innovation involved in the formulation of the new concept of pressure, and the long historical path to that accomplishment, have gone largely unheeded in the existing literature. In rectifying this situation I hope to shed new

[1] The common sense distinction includes the recognition that wine, milk and a range of other liquids differ from solids in the same way that water does, insofar as they can flow, spill and so on. So while hydrostatics typically involved dealings with water, as the 'hydro' implies, it was generally assumed that it applies to other liquids also. Consequently, in this book I move freely and without qualification between talk of water and talk of liquids in general, as, indeed, the historical actors in our story did.

A.F. Chalmers, *One Hundred Years of Pressure*, Archimedes 51,
DOI 10.1007/978-3-319-56529-3_1

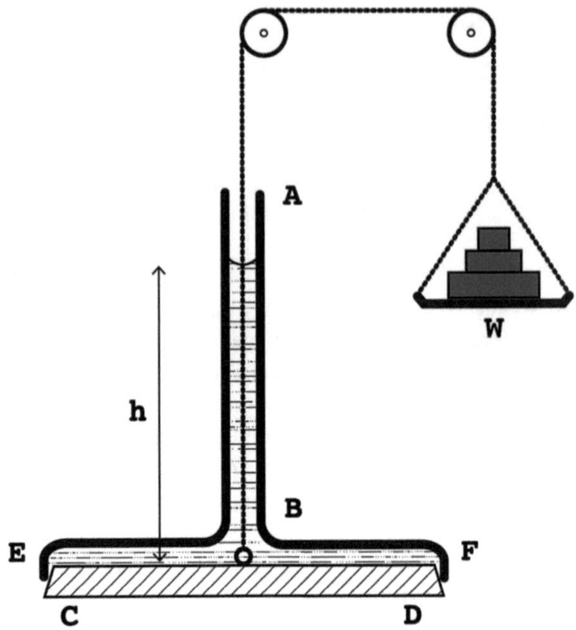

Fig. 1.1 It can take several hundreds of pound to support an ounce of water

light on the nature of some of the novelties that emerged in science in the period that separates the work of a young Nicholas Copernicus from that of a mature Isaac Newton which is commonly referred to as the Scientific Revolution.[2]

In this opening chapter I endeavor to convince the reader that the story of the emergence of hydrostatics is intellectually exciting. To achieve this end I dramatize the situation somewhat. Yet I am mindful of the fact that in the course of later discussion I will need to be more precise and more circumspect. In this opening chapter I freely view the past from the point of view of the present, alert to the fact that this must not be done in a way that prevents us from learning what there is to be learnt from that past and the path that brought us from there to the present. I make some bold historical claims that are transformed into precise, well-documented and

[2]The practice of science in the age of Newton differed markedly from its practice in the age of Copernicus. I take the term 'the Scientific Revolution' to refer to the sum of the changes that took place in that period without any presumption that the various novelties formed a coherent whole. In this respect I follow David Wootton (2015) in his recent book, *The Invention of Science*. I doubt that there is a conception of a 'scientific revolution' that is sufficiently precise or cogent to make a debate about whether there was or wasn't one meaningful or fruitful but this leaves plenty of room for debate about the sense in which various specific changes involved continuities and discontinuities. Among the significant changes that took place in science in the period in question are those that relate to the role of experiment and mathematics and to some fundamental conceptual changes. It is by focusing on these issues in the context of seventeenth-century hydrostatics that I aim to shed light on some, but not all, of the key changes generally regarded as important ones occurring in the period referred to as 'the Scientific Revolution'.

substantiated claims in the course of the book. I presume the story to be a novel and informative one prior to demonstrating as much in the following pages.

1.2 Some Puzzles Posed by Liquids in Equilibrium

Moving beyond the common distinction between liquids and solids to a comprehensive and precise theorization of it is far from obvious. In this section I point to some situations involving liquids at rest that can prove puzzling even for a modern reader. This must have been the case so much more for a scholar in the late sixteenth century, which is where our detailed story of the emergence of hydrostatics begins.

I am confident that it will come as a surprise to many readers that, in the situation like the one depicted in Fig. 1.1, it can take a weight, W, of many pounds to prevent the small amount of water confined in the region ABFE from forcing out the stopper CD, even if the weight of that water is less than a few ounces.

It has been known since antiquity that water supplied to a town from a lake in the nearby hills is unable to rise naturally above the level of the water in the lake. However, suppose we consider a situation where water from the lake rests in equilibrium with water in a pipe at that maximum height. How can it be that the relatively small amount of water in the pipe can support the water in the entire lake that supplies and weighs down on it?

Both water and sand can be transferred from a high to a low altitude via a pipe. However, water will pass freely around a bend in the pipe that causes the water to interrupt its downward journey by a small ascent but sand will not. If water in one limb of a U-tube is pressed downwards the water level in the other limb will rise. The same does not happen if the water is replaced by sand. How are we to comprehend these differences in behavior?

Water presses against a lock gate, as evidenced by the difficulty encountered when attempting to open it against a head of water. How can this be reconciled with the presumption that the weight of water presses vertically *downward* and that the horizontal component of weight is zero? How can it be that the water trapped between two lock gates presses equally on those gates whether they are separated by a few millimeters or by several kilometers, provided that the head of water is the same in each case?

Those of my readers who lack a background in physics, and quite a few of those who do not, will very likely be challenged by at least some of my examples. This is so in spite of the circumstance that a theory of hydrostatics capable of solving the puzzles I have posed has been available since late in the seventeenth century. The theorization of the difference between solids and liquids supplied by hydrostatics is far from obvious, even to many educated people of today.

All of my examples were known towards the end of the sixteenth century. Simon Stevin, in 1586, made the first attempt to formulate a theory of hydrostatics that moved beyond what Archimedes had set in place. Stevin was involved in the design of lock gates and explicitly pinpointed the facts about them that I have problematized above. Figure 1.1 is my version of one that was depicted by Stevin himself (Dijksterhuis 1955, p. 488). Stevin was in a position to appreciate and note many facts about the behavior of liquids but his theory fell short of adequately theorizing them, as we shall see. It took a century to arrive at the point where Newton more or less accomplished that.

1.3 Understanding Liquids via a Technical Sense of Pressure

Solid objects cohere as wholes. While their shapes can be distorted to some degree to the extent that they are elastic, in the main they accelerate or tend to move as wholes in the direction of an applied force. Bodies of liquid differ in this respect. They freely adapt their shape to a containing vessel or an immersed solid and will simply give way to a solid object moving slowly through them. If a liquid body is subject to a force it will not tend to move as a whole in the direction of that force. Rather, the applied force is converted by the liquid into an isotropic one that will urge the liquid to move in any direction open to it, unconstrained by any tendency for the liquid body to cohere. Water will tend to leak from a pipe with equal facility in any direction and not just in the direction of the weight of the head of water bearing down on it.

To meet the demands of a theory of hydrostatics it proved necessary to give a precise and quantitative formulation of the features of liquids described in the previous paragraph. If a body of liquid is pressed at some region on its surface, then that pressing, measured as a force per unit area, is communicated through the body of the liquid, acting on any surface within it with the same force per unit area whatever the orientation of that surface. The forces acting on either side of any surface within the body of a liquid at equilibrium are equal and opposite, so the net force is zero. It is only where the liquid is in contact with a solid surface that the pressing is not balanced by a counter-pressing and results in a net force on the solid surface.

The above two paragraphs contain the beginnings of an exposition of a technical sense of pressure, a sense that is taken for granted in post seventeenth-century physics. A distinctive feature of the history of hydrostatics to be presented in the following pages stems from the importance I attribute to the distinction between the technical sense of pressure, alluded to above, and a common sense of pressure dating from ancient times and referred to by the Latin *pressio/pressionis*. Seventeenth-century advances in hydrostatics necessitated replacing the common by a technical concept of pressure. This is what lies behind the main title of the book. It will help to clarify at the outset the distinction between the two concepts of pressure.

The technical concept denotes the stress on a plane within a fluid at any given location, measured as a force per unit area.[3] In spite of its specification as a force per

[3] The modern concept of fluid includes gases as well as liquids. It was not until the mid-seventeenth century that the theorization of gases, or rather, air, as distinct from liquids, became a serious issue. Prior to that, talk of 'fluids', by Archimedes for example, amounted to talk of liquids. In the main

unit area, pressure, unlike force, is not a directed quantity. It is a *scalar*, a property specified solely by its magnitude, as opposed to a *vector*, a property having a direction as well as a magnitude. In view of this, the equation $\underline{F} = P.A$, giving the force due to pressure P on surface of area A, can be puzzling insofar as the left hand side of the equation is expressed as a vector whereas the right-hand side appears as the product of two scalars. The puzzle is resolved once it is realized that an area of surface, A, is directed like a vector insofar as it has an orientation. The orientation of any particular surface can be specified by a unit vector, $\underline{i}$, normal to the surface. We can then write the expression for the force on a surface as $\underline{F} = P.A.\underline{i}$, an expression that now equates two vectors rather than a vector and a scalar. In a fluid the force due to pressure on a plane of area A is normal to it and independent of its orientation. Pressure within a fluid in equilibrium is isotropic. The gas equation, PV = nRT that became central to thermodynamics in the nineteenth century relates pressure to two other scalars, volume and temperature. Within the body of a fluid the net force due to pressure, which acts equally in all directions, is zero. It is variations in pressure, such as those that occur at the boundary between a fluid and a solid, that lead to forces.

The key property of fluids that distinguishes them from solids is that the former are unable to sustain a tangential stress. The force on any surface within the body of a fluid in equilibrium is normal to that surface and its magnitude per unit area is independent of the orientation of the surface. I do not suggest for a moment that all this is obvious. It was not until Newton, or perhaps Euler, that this was fully understood. That is precisely why the story I have to tell spans a century and why we should be suspicious of any history of hydrostatics that takes pressure as a given.

Let us now follow this reflection on the concept of pressure as it occurs in modern physics with similar reflection on a common conception of pressure employed in ancient times and which persists in today's everyday discourse. Many common usages to be found in Latin texts involve the effect of the weight of solids. The flattening of the grass beneath a blanket at a picnic is attributed to the pressure caused by the weights upon it. An example that moves beyond weight involves the expansion of a bladder caused by forcing a fluid into it. It is natural to understand the expansion as the result of pressure, and the bursting of the bladder as a result of that pressure becoming more than the bladder can withstand. These common usages of the term 'pressure' all involve forces on a solid surface. The verb 'to press', *premo* in Latin, carries with it the idea of a directed force. Individual examples of pressing have a direction and as such are vectors. The weighing of solids against surfaces constitutes a paradigmatic example of pressure in its common sense, as implicit in the use of the example involving the flattening of grass at a picnic. It still makes common sense to note that the floor of a loft must be built so that it is sufficiently strong to bear the *pressure* of the load it is intended to carry, the weight per unit area rather than the total weight. An explication of why it matters which way round a drawing pin is orientated involves pressure in its common sense.

---

it is clear from the context whether 'fluid' is to be interpreted as referring to liquids or in the more inclusive sense and I explicitly raise the issue of the distinction only where it is necessary.

A key difference between pressure in the common sense and pressure in the technical sense is that the former relates to forces on bounding surfaces between media whereas the latter refers to forces within the body of media. Another is that, from the technical point of view, pressure is a scalar not a vector. Directed forces, such as those that occur at the boundary of a liquid are determined by variations of pressure, the gradient of pressure in technical terms, rather than by pressure itself. The technical concept of pressure in fluids breaks from the directedness implicit in the verb 'to press' from which 'pressure' originally derived and which is presupposed in the concept of pressure in its common sense.

I do not introduce the contrast between the two notions of pressure in order to cast judgment on the seventeenth century contributors to hydrostatics according to how far they moved beyond the common to the technical sense of pressure. Little by way of historical understanding would be achieved by doing that. Nor do I invoke the modern sense of pressure to serve an explanatory function, as if it were acting as an attractor drawing the progress of hydrostatics towards it. My purpose is to highlight the fact that if we are to understand the notion of pressure as it occurs in seventeenth-century hydrostatics, then there is work to be done to discern exactly what was intended by it and related terms in each instance of their use. The literature on the history of hydrostatics suffers for want of such work.[4] The norm is for the concept of pressure to be taken as a given and not in need of comment or analysis, thereby leaving the way open for an interpretation of the historical texts that presumes that 'pressure' is employed in them in something like its modern sense. This leaves the crucial changes in and development of the notion of pressure unrecognized. I stress the distinction between the common and technical senses of pressure at the start of my study so that we are well prepared to avoid the misleading history that can arise by ignoring it.

Insensitivity to developments in the concept of pressure of the kind that I am highlighting is apparent in translations into English of seventeenth century texts. Translations of hydrostatic works by Simon Stevin in Dutch and by Gilles Personne de Roberval and Blaise Pascal in French make free use of the term 'pressure' without comment, presumably in the expectation that the reader will interpret it in something like its modern sense.[5] The problem is exacerbated by the fact that, while the original texts talk of pressing, they rarely use a noun (*drucking* in Stevin's Dutch, *la pression* in French) that is the equivalent of 'pressure'.[6]

---

[4] Pierre Duhem is one of the few historians to have given serious attention to some of the nuances involved with respect to the concept of pressure. See, for instance, Duhem (1905).

[5] The works in question are the translation of Stevin's hydrostatics in Dijksterhuis (1955), the translation of Pascal's treatises on hydrostatics and pneumatics in Pascal (1937) and the translation of the relevant letters of Roberval in Webster (1965).

[6] The influential analysis of Boyle's pneumatics by Steven Shapin and Simon Schaffer involves confusions of their own making because of the inadequate attention the authors pay to the precise meaning of 'pressure' as it evolved in Boyle's writings. The confusions occur in Shapin and Schaffer (1985), especially Chap. 2. These matters are discussed in Chap. 8, footnote 13.

1.4 Hydrostatics as an Extension of the Statics of Simple Machines

There are good historical reasons for thinking of early attempts at a theory of hydrostatics as extensions of the mechanics of simple machines: the lever, pulley, wedge, wheel and axle and screw. Towards the end of the sixteenth century a theory of statics in the style of Archimedes had been developed with considerable sophistication.[7] Not only could its content be taken for granted by a scholar seeking to venture into hydrostatics, but it could also be taken as an exemplar of what a mathematized physical theory should be like. Stevin presented his *Elements of Hydrostatics* as a sequel to his own theory of statics, *The Art of Weighing*, both published in 1586. However, the extension of statics to include the equilibrium of systems involving liquids posed a challenge that was not fully appreciated as such by the historical actors. It has not been adequately appreciated by subsequent historians either.

I suggest that the reason why there could be a theory of statics prior to the Scientific Revolution is similar to why there could be and was a theorized geometry. Concepts and propositions sufficient to serve as the foundations of geometry were idealized notions of what could be readily abstracted from common experience involving such things as plumb lines and lines of sight. The concept of weight that was central to the understanding of simple machines can likewise be seen as an abstraction from common knowledge implicit in age-old practices. A material object has weight insofar as it presses downward, where the term 'downward' needs to be interpreted as 'towards the center of the earth' once the spherical shape of the earth is appreciated. The weight of a combination of objects is equal to the sum of their individual weight. Weights can be compared using a balance.

There is more to statics than an appreciation and measurement of weight. It is also necessary to comprehend the constraints responsible for the workings of the various machines. The theory of the balance presumes that a balance arm is rigid and constrained to rotate about its point of suspension. Other machines involve the constant length of pulley strings, the rigidity of gear wheels and the like. Idealizations of such constraints suitable for at least the beginnings of a science of statics could, like weight, be abstracted from what was implicit in common, practical knowledge with a long history.[8] In this respect it is relevant to note that when assumptions about the constraints at work in the operation of simple machines, such as the rigidity of balance arms, were employed in the science of weight, this was done implicitly and without comment.

A shared feature of the constraints at work in simple machines is that they all involve solidity in one form or another. None of them can be applied to liquids. If the theory of simple machines is to be extended to hydrostatics then the challenge

[7] For an indication of the degree of sophistication reached see Renn and Damerow (2012).

[8] This is not to deny that sixteenth century statics was developed with a degree of sophistication that transcended common sense and the knowledge of practitioners, as is clearly evident from the work referred to in footnote 7.

that must be met can be expressed as follows: What mechanical constraints are imposed on a material system by the presence of a body of liquid? The history of hydrostatics, as well as any discomfiture that might have been engendered in the modern reader by the puzzles outlined in Sect. 1.2, attests to the fact that the answer to this question was and is far from obvious. A notion of solidity sufficient for an understanding of the rigidity of balance arms, the pushing of one gear wheel by another and the transmission of tension via a cord wrapped around pulleys could be readily abstracted from common knowledge. The same could not be said for the mechanical action of liquids. As far as the latter is concerned, late sixteenth-century familiarity with liquids could lead so far, but not far enough.

Eighty years after Stevin's attempt to formalize a science of hydrostatics we find a young Newton recognizing that what was needed for hydrostatics was an appropriate definition of the liquid state that would serve to identify the mechanical action of liquids. By the time he wrote the *Principia* he had cleared up his early muddles and provided an adequate definition. Armed with it, Newton was able to develop a version of hydrostatics that came close to that which is accepted today. Newton succeeded where others had failed because he was able to pinpoint the precise way in which liquids constrain a mechanical system. Some readers may struggle to fully comprehend his solution, described and analyzed in Chap. 9, which makes crucial use of a technical concept of pressure. The move from statics to hydrostatics involved significant conceptual innovation that was gradually and painstakingly accomplished in the years that separated Stevin and Newton and has been inadequately grasped by historians hitherto.

1.5 Hydrostatics and the Scientific Revolution: Experiment and Mathematics

In the history that follows I aim to document the way in which a science of hydrostatics emerged in the seventeenth century largely as the result of the development of a concept of pressure up to the task. I identify the claims made and the ways in which they were defended and deployed by the main historical actors from Stevin to Newton. In this respect I offer what might be termed a descriptive history of science, defending my descriptions as the correct ones by reference to the relevant historical sources. But that is not all I aim to do. I do not present my descriptive history as an end in itself but as a means of throwing light on the nature of science and its origins in the Scientific Revolution. It is typically assumed that novel characteristics of the emergent sciences were the role played by experiment and the degree to which it was mathematized. Here the words of the historical actors need not be accorded the authority they need to be given in the case of descriptive history. My diagnoses of the ways in which experiment and mathematics were deployed in the various versions of hydrostatics are to be judged by the extent to which they are borne out by and make sense of the descriptive history. I can legitimately aim to do better than the historical actors did in this respect.

A focus on experimentation led some participants in the revolution to refer to the new approach as 'the experimental philosophy'. A focus on mathematization was implicit in Newton's description of his new system of dynamics as the 'mathematical principles of natural philosophy'. In this Section I foreshadow the way in which this study of hydrostatics will throw some light on the subtleties involved in the relationship between the experimental and mathematical dimensions of seventeenth-century science.

Thinking of hydrostatics as an extension of statics helps to pinpoint a puzzle the solution to which is far from straightforward, requiring some circumspect thought about the relationship between theory and experiment. From a modern perspective the issue can be posed as whether hydrostatics is to be classified as mathematics or physics. Around the turn of the sixteenth century it was the norm to treat statics as a theory deduced from unproblematic axioms and as akin to geometry in that respect. From such a perspective it is as wrong-headed to suppose that statics needs to be tested by experiment as to suppose that Pythagoras's theorem is to be supported by measuring the sides of material triangles. From this point of view, the deduction of the balance conditions for a bent lever, say, is to be regarded as an application of the theory of statics rather than as providing an opportunity to test or confirm it. Stevin made it quite explicit that this was the way he interpreted both his statics and his hydrostatics, as we shall see. But does not this imply that statics and hydrostatics are mathematics in the sense that geometry is, with the consequence that experiment plays no role other than that of practical application?

In Chalmers (2015) I compared the hydrostatics of Stevin with that of Pascal written in the mid-seventeenth century and suggested a response to the above conundrum. In effect, I defended the position that the move from the statics of Stevin to that of Pascal corresponded to a shift from hydrostatics as mathematics to hydrostatics as experimental science. Whereas Stevin sought a theory consisting of propositions deduced from unproblematic axioms, Pascal construed his theory as one that was to be acceptable to the degree that it could be supported by experiment. Explicit remarks by Stevin and Pascal support such an interpretation. So Stevin's treatment of hydrostatics as akin to geometry represented the old way and Pascal's emphasis on the need for experimental support for hydrostatics represented the new way!

My subsequent engagement with Newton's hydrostatics forced me to seriously rethink that interpretation. On the one hand, I came to recognize that Newton's hydrostatics amounted to a successful theorization of the liquid state and as such marked the culmination of the story I had set out to tell. On the other hand, I had to recognize that when Newton explicitly posed the contrast between the mathematical method, similar to that followed by Stevin, and the experimental method, similar to that advocated by Pascal, *he sided with the former*. Newton presented his hydrostatics as following from his definition of a liquid. According to him, phenomena produced via experiment play a subsidiary and incidental role as far as the vindication of his hydrostatics is concerned.

Detailed treatment of this conundrum cannot be fully dealt with in an introduction. I raise it here as a foretaste of the ways in which, in the ensuing history of hydrostatics, I grapple with fundamental issues concerning the roles of experiment and mathematics in science and the ways in which they were transformed in the seventeenth century. I will find it necessary to consider roles played by experiment other than providing the evidence against which the merit of a theory is to be judged. I endeavor to utilize my history of hydrostatics to throw light on the kind of knowledge that science is, distinct from practical know-how and the conceptualizations necessary for it and also from metaphysics as embodied in the world-views of the Ancient Greek philosophers, scholastic Aristotelians and mechanical philosophers of the seventeenth century.

1.6 Mechanism and the Mechanical Philosophy

I was motivated to carry out this historical study of the emergence of hydrostatics as a response to criticisms of my book, *The Scientist's Atom and the Philosopher's Stone* (Chalmers 2009). In that work I drew a distinction between two kinds of theories about atoms, philosophical versions that were essentially accounts of the ultimate structure of material reality and scientific versions that arose in response to specific problems within the sciences. What proved to be the most controversial part of my book was the application of my thesis to the seventeenth century. I argued that atomism as it was articulated by mechanical philosophers such as Gassendi and Boyle did not feed productively into science and, further, that scientific versions of atomism that had developed by the late nineteenth century owed nothing to it.

The issue I raise in this Section is largely irrelevant to the historical story in this book. I raise it nevertheless because its irrelevance is highly relevant! The issue concerns the relationship between the development of hydrostatics and other sciences in the seventeenth century that were modeled on the science of simple machines, on the one hand, and the change in world-view, from an Aristotelian to a mechanical one, on the other. The latter change corresponds to what E.J. Dijksterhuis famously described as the 'mechanization of the world picture'.[9] Both the new sciences and the new world-view were mechanical, but they were mechanical in senses that need to be carefully distinguished.

A central element of the world-view that came to be referred to as the mechanical philosophy was the account of the fundamental make-up of the material world held by its proponents. According to that account, at rock bottom the world consists of particles of universal matter characterized by their shape, size and motion. The forms presumed to animate the world according to scholastic Aristotelians were banished by the mechanical philosophers on the grounds that their mode of existence is unintelligible insofar as they are other than matter. The details of the mechanical philosophy differed from one mechanical-philosopher to another.

[9] See Dijksterhuis (1969).

Boyle's version was largely confined to a defense of the ontological thesis to which I have here drawn attention. Rene Descartes, by contrast, was not content with the ontological thesis alone but aspired to add the broad details of the corpuscular mechanisms that constitute our universe. Such details aside, the ontological thesis that the material world was to be reduced to particles of matter characterized solely by their shape, size and motion was central to the mechanical philosophy. The key argument for it was that it was intelligible in a way that the Aristotelian alternative was not.

The simple machines, the theorization of which constituted the model for much seventeenth-century science, were certainly mechanical. Indeed, they were paradigmatic mechanisms. However, they differed from corpuscular mechanisms in a number of significant respects and an appreciation of this is crucial for understanding developments in seventeenth-century science.

Simple machines can be understood in terms of their parts, the properties of those parts and the way in which the parts are connected together to form the whole machine. The functioning of a machine as a machine is comprehended in terms of the way in which cause is connected to effect by way of the material connection of the parts. A balance consists of a balance arm supported by a pivot and weights in scale pans suspended from the arm. The weight acting down at the extremity of one arm is transmitted via the rigid balance arm to the counterweight at the other extremity, the latter being urged upward as a result. All this was commonly assumed in dealings with simple machines.

It is enlightening to consider how the notion of mechanism that I have illustrated by reference to a simple balance was extended to applications in seventeenth-century anatomy.[10] Many anatomists came to think of organisms as machines consisting of parts that were connected together in a specific way. From this point of view, understanding an organism required the identification of the parts, such as muscles or nerves, the properties of those parts, such as the elasticity of muscles and the ability of nerves to transmit signals, and the way in which those parts were connected together in an organism. Investigation of such matters involved mechanical interventions such as dissections and an analysis of the properties of the parts considered separately as well as the way in which they were organized as a whole.

If we now look back at simple machines from the perspective of the kind of anatomy identified above, then features of them stand out in a way that might otherwise go unheeded. Understanding machines requires an understanding of the properties of their parts. However, in the context of the simple machines such properties are so obvious as to be hardly worthy of mention. The facts that a balance arm is rigid and that weights act vertically downwards are cases in point. The constraints at work in simple machines are in effect their defining characteristics but were too obvious to be singled out and identified as assumptions.

It is instructive to bring the perspective described above to bear on hydrostatics. How are we to understand machines that include a body of liquid as a key component? An answer to this question requires that the properties of liquids as such be

[10] Here I draw on the work of Bertoloni Meli (2011). See especially pp. 12–16.

understood to the extent that the constraints they impose on a material system can be identified. The answer to the question proved to be far from obvious and, as it transpired, required a technical notion of pressure that was not available at the beginning of the seventeenth century.

We are now in a position to spell out key differences between machines, including organic 'machines', and corpuscular mechanisms as they figured in the mechanical philosophy. Mechanical philosophers sought to identify the ultimate structure of the material world. Their answer involved appeal to ultimate particles possessing shape, size and motion. Those properties were ultimate insofar as they were treated as not in need of explanation at a deeper level, and they were specified in advance of any empirical investigation. By contrast, the properties of the parts of machines needed to be established as the result of empirical or experimental investigation and an adequate specification of them did not require that they be explained at some deeper, let alone ultimate, level. Properties such as the rigidity of balance arms or the elasticity of muscles are what Boyle was to call intermediate causes as opposed to the ultimate causes that mechanical philosophers sought to locate in the shapes, sizes and motions of corpuscles of universal matter.[11]

The historical path in statics from Aristotle and Archimedes to the beginning of the seventeenth-century involved mechanical considerations of the way in which machines functioned as a result of the combination of their parts characterized in terms of their (intermediate) properties. Hydrostatics arose as an attempt to extend that program to machines involving bodies of water. Concerns about how to characterize the ultimate structure of material reality were simply not involved. The situation changed in the seventeenth century once the mechanical philosophy had been articulated as an alternative to an Aristotelian world-view. At that stage the relationship between the two kinds of mechanism was an issue that had a bearing on philosophical questions of pressing concern at the time. The issue will come to the fore in our history of hydrostatics when we consider Boyle and Descartes, both of whom contributed to the new science and to an articulation of versions of the mechanical philosophy. I argue the case that the relationship between hydrostatics, alongside other 'mechanical' sciences, and the mechanical philosophy was not of a kind that enabled the former to have much to gain from the latter.

While the project of explaining hydrostatic effects by appeal to intermediate causes was low level compared to the ambitious attempts by the Ancient Greeks and the seventeenth-century mechanical philosophers to comprehend the ultimate make-up and structure of the Universe, it nevertheless went beyond the practical knowledge of artisans. Seventeenth-century hydrostatics was expected to explain the range of hydrostatic effects by identifying their efficient causes. Early attempts to do so appealed to a range of properties of liquids, such as weight and some common understanding of their capacity to flow or support floating, in an opportunist and ad hoc way. Advances and improvements eventually culminated in Newton's hydrostatics, which was derived from a definition of the liquid state and was sufficient to explain and co-ordinate known hydrostatic phenomena. The articulation of the

[11] Boyle's views on mechanism and intermediate causes are discussed in detail in Sect. 8.3.

definition and its consequences required a technical concept of pressure. This new concept was low-level compared to the ultimate causes sought by mechanical philosophers. But it was not obvious or mundane. Its fabrication came as the result of a century of intellectual endeavor. The new hydrostatics made possible an understanding of the behavior of liquids as distinct from solids that had not been available a century earlier. As far as practical know-how is concerned, the new understanding did not yield immediate practical benefit.

Using unrefined categories of modern times, the considerations of the previous paragraph can be crudely summarized as follows: The seventeenth century witnessed the emergence of hydrostatics as a science, as distinct from both philosophy and practitioners' knowledge. As I have mentioned, the defense of this thesis gave me a motive for studying the history of hydrostatics. In the course of my study I came to appreciate that it needed to be understood in terms of the gradual formation of a technical notion of pressure. The details of that story can be appreciated and evaluated independently of my claims about the importance of the distinction between ultimate and intermediate causes to the extent that I could have written a descriptive history of hydrostatics that largely avoided mention of the distinction. I have not done that because I believe the issue has an important bearing on the nature of science and the changes that its emergence in the seventeenth century involved.

1.7 The Structure of the Book

The first systematic treatise of the modern era on hydrostatics was written by Stevin in 1586. Following a sketch of the relevant historical background in Chap. 2, including the work of Archimedes, a detailed exposition and analysis of Stevin's theory is given in Chap. 3. I identify limitations of his theory when it is considered as a theorization of the distinction between solids and liquids. It was three quarters of a century before Pascal and then Boyle took the theory of hydrostatics beyond Stevin's efforts by making significant moves towards refining the concept of pressure. This paved the way for Newton to formulate a version of hydrostatics that came close to what is accepted today and which represents the culmination of my historical story. The contributions of Pascal, Boyle and Newton are discussed in Chaps. 6, 8 and 9 respectively.

The story of the progression from Archimedes to Newton is interrupted by two digressions. Chapters. 4 and 5 deal respectively with the contributions of Galileo and Descartes to hydrostatics. Neither of those great minds made significant advances towards the concept of pressure that proved to be necessary for a theorization of the liquid state. This serves to emphasize the extent to which those advances were far from obvious at the time.

Chapter 7 involves the second necessary digression. It is concerned with the advances in pneumatics that took place in the wake of Torricelli's famous experiment which have considerable significance for our story. Those advances brought to the fore the need to capture the natural tendency of air to expand by attributing to it a property that was distinct from and an addition to weight, the property that Boyle referred to as its spring. Since liquids do not have a significant spring, the question of what distinguishes them from solids and air came to the fore. It was against this background that Boyle made some important conceptual moves towards a technical concept of pressure in addition to his better-known experimental contributions.

In Chap. 10 I summarize and stress the significance of the steps in the path to a novel and technical sense of pressure, extracting the main points from the historical detail. In Chap. 11 I reflect on the role of experiment in hydrostatics and attempt to capture the sense in which it was an empirical science in spite of the fact that novel experimental support for it did not play a significant role in the seventeenth-century advances. In the short, concluding chapter I pinpoint the main lessons to be learnt from my history of hydrostatics as just one component of the significant, though diverse, changes that constituted the Scientific Revolution.

References

Bertoloni Meli, D. 2011. *Mechanism, experiment, disease: Marcello Malpighi and seventeenth-century anatomy*. Baltimore: The John Hopkins University Press.

Chalmers, A. 2009. *The scientist's atom and the philosopher's stone: How science succeeded and philosophy failed to gain knowledge of atoms*. Dordrecht: Springer.

———. 2015. Qualitative novelty in seventeenth-century science: Hydrostatics from Stevin to Pascal. *Studies in History and Philosophy of Science*. 51: 1–10.

Dijksterhuis, E.J. 1955. *The principal works of Simon Stevin, Volume 1, Mechanics*. Amsterdam: Swets and Zeitlinger.

———. 1969. *The mechanization of the world picture*. Oxford: Oxford University Press.

Duhem, P. 1905. Le Principe de Pascal. *Revue Générale des Sciences Pures et Applique*. 16: 599–610.

Pascal, B. 1937. *The physical treatises of Pascal: The equilibrium of liquids and the weight of the mass of the air.* Trans. A. G. H. Spiers and I. H. B. Spiers. New York: Columbia University Press.

Renn, J., and P. Damerow. 2012. *The equilibrium controversy: Guidobaldo del Monte's critical notes on the mechanics of Jordanus and Benedetti and their historical and conceptual background, Max Planck Research Library for the history and development of knowledge. Sources 2*. Berlin: Edition Open Access. http://www.edition-open-access.de.

Shapin, S., and S. Schaffer. 1985. *Leviathan and the air pump: Hobbes, Boyle and the experimental life*. Princeton: Princeton University Press.

Webster, C. 1965. The discovery of Boyle's law and the concept of the elasticity of air in the seventeenth century. *Archive for History of Exact Sciences*. 2: 441–499.

Wootton, D. 2015. *The invention of science: A new history of the scientific revolution*. London: Penguin Random House.

Chapter 2
The Historical Background to Stevin's Hydrostatics

Abstract The starting point of my story is the account of hydrostatics formulated by Simon Stevin in 1586. He could take for granted a distinction between liquids and solids which formed part of everyday knowledge and which was presumed in common technologies involving the management of water. Stevin could also take for granted the science of simple machines that had been developed with considerable sophistication in the sixteenth century. Stevin explicitly based his theory of hydrostatics on Archimedes' work on floating bodies, building on its content as well as adopting its style of reasoning. The propositions of hydrostatics were to be derived from postulates sufficiently unproblematic to be granted at the outset.

Our detailed story of the emergence of a science of hydrostatics begins with Simon Stevin's *Elements of Hydrostatics*, published in 1586. In this chapter the scene is set for an appraisal of that text by giving an outline of the resources that Stevin drew on in the construction of his theory. Those resources included everyday familiarity with the differences between solids, liquids and air and a range of technologies that utilized the natural properties of water. They also included various attempts to theorize weight and motion that had origins in the work of the Ancients and underwent significant developments in the medieval and Renaissance periods. Finally, Stevin built on the one significant attempt to theorize hydrostatics that preceded his own, Archimedes' book on floating bodies.

2.1 Common Sense and Technological Knowledge of Liquids as Distinct from Solids

Implicit in hydrostatics, insofar as it is a science of the equilibrium of *liquids,* is a formalized characterization of the nature of the liquid state. By the time we reach Isaac Newton we will find him setting out his science of hydrostatics as following from a definition of liquids. However, an appreciation of some distinction between solids, liquids and air did not require a formal theorization of it. Stevin drew on everyday knowledge of the distinctive properties of liquids which is implicit in and

A.F. Chalmers, *One Hundred Years of Pressure*, Archimedes 51,
DOI 10.1007/978-3-319-56529-3_2

necessary for everyday life, as did the other figures in our story from Archimedes onwards. Liquids can spill in a way that solids cannot, which is why the former, unlike the latter, need to be stored in a solid container. Solid objects retain a distinctive shape whereas samples of liquid adapt their shape to a container or spread over a surface if unconstrained by a container. Liquids can be drunk, unlike solids that need to be chewed before they are swallowed. Solids like wood can float in water, and a raft made of logs will sink deeper the more weight it carries. It is possible to swim in water but not in ice, and to skate on ice but not on water. Water freely yields to a hand moving slowly through it whereas a solid impedes any such motion.

Human survival would hardly be possible without primitive knowledge about liquids of the kind described above. But such survival was soon facilitated by technologies that built on, exploited and refined this primitive knowledge. Sophisticated technologies for irrigation and the reclamation of land predated the rational theorizing about the world introduced by the Ancient Greeks by several thousand years. The development of those technologies in Egypt and Mesopotamia involved extensive systems of canals and reservoirs and the use of dykes and dams. The transport of water via conduits and aqueducts dates from 1000 BC and lock gates were introduced in China around the same time. Wells were constructed and water scooped from them by buckets moving on endless chains. The syphon, the pipette, the syringe and bellows for fanning fires are devices that were all known to the Ancients.

It is possible to infer from details of such technologies that considerable knowledge about liquids was implicit in them whether this was made explicit or not at the time of their use. It became evident, for example, that water conveyed to a city via conduits could not flow naturally to locations that were higher than the level of the water in the lake or reservoir.[1] It also became evident that pipes conveying water from a high source needed to be strong enough to withstand the pressure.[2]

The extent to which practical knowledge developed at a particular time and location was known and employed at later times and other geographical locations was largely a matter of historical contingency. But to the extent that practical knowledge was passed on, its progress was cumulative insofar as one generation could take for granted the state of practical arts they inherited and were able to build on and improve them. Both points can be illustrated by an example very relevant for our history of hydrostatics. Lock gates, which can be used to raise and lower vessels and reclaim land, seem to have been known to the Chinese some 2000 years before they were first introduced into Europe, in the Netherlands in 1373. About a century later their use was facilitated via an innovation introduced by Leonardo da Vinci in 1487.

[1]The Roman engineer Sextus Julius Frontius (40–103 A.D.) made this point explicit. 'The several aqueducts reach the city at different elevations. Whence it comes about that some deliver water on higher grounds, while others cannot elevate themselves to the higher summits.' (Cited in Rouse and Ince 1963, p. 27).

[2]Marcus Vitruvius (first century BC) addressed the problem when he gave careful instructions about how to make conduits by bending lead sheets into a circle, paying special attention to the thickness of the lead sheet and the making of the seal by butting the edges of the bent sheet and pouring molten lead along the joint. (Rouse and Ince 1963, p. 28).

It involved V-shaped lock gates pointing upstream, which had the advantage that the water pressing horizontally on the gates served to maintain the seal.

Knowledge of the distinctive behavior of air, as distinct from solids and liquids, was implicit in everyday phenomena such as breathing and winds and in technologies involving sailing ships, bellows and windmills.

2.2 The Science of Simple Machines: Statics

By the time Stevin began his work on mechanics and hydrostatics, the works of Archimedes on centers of gravity and on floating bodies, as well as the Aristotelian treatment of the lever, had been freely available in Western Europe in Latin translation for several decades.[3] Stevin's *The Art of Weighing* was modeled on and an extension of Archimedes' two books *On the Equilibrium of Planes*. A key notion was that of the center of gravity of a body, the point through which its weight acts. Archimedes deployed this notion in order to comprehend the equilibrium of the balance with equal arms and extended it to a treatment of a balance with unequal arms. He determined the position of the centers of gravity of a range of plane geometrical figures.

The preceding sentence serves to highlight the extent to which Archimedes' statics involved idealizations of physical situations. Balances were depicted geometrically, with balance arms represented as lines or thin rectangles and weights represented as plane figures with simple geometrical shapes. Postulates with physical content, such as Postulate 1, 'weights that are in equilibrium at cqual distances are equal' and Postulate 3, 'Unequal weights can only be in equilibrium at unequal distances, the greater weight being at the lesser distance' are discussed in relation to idealized depictions of balances represented by simple plane figures. At a superficial level at least, there is a sense in which Archimedes did prove theorems mathematically which nevertheless had empirical content. Real balances with unequal arms do exhibit balance conditions in approximate conformity with Archimedes' mathematical propositions, with the 'approximate' here reflecting the fact that the theoretical treatment involves idealizations of messier real world situations. The tensions or problems involved here sometimes surfaced as explicit challenges, as clearly exhibited in the Equilibrium Controversy of the Sixteenth Century. It involved conflicting views on whether or not it is legitimate to ignore the fact that lines supporting the weights on either side of a balance are not strictly parallel and

[3] A pioneering work on the history of statics is Duhem (1991), an English translation of the French original published in 1905/6, with an introduction by Stanley L. Jaki. Other dated but still useful overviews are Clagett (1959), Moody and Clagett (1960) and Drake and Drabkin (1969). More recent studies which serve as an introduction to the contemporary debate are Palmieri (2008) and Van Leeuven (2016). Details of the historical transmission of developments in theoretical mechanics of which Stevin was in a position to take advantage are given in Renn and Damerow (2012, Chapter 3).This latter work serves to illustrate the degree of sophistication reached within the science of weight towards the end of the sixteenth century.

whether it is legitimate to ignore the finite dimensions of the balance arm. We will soon be confronting the specific challenges posed by the extension of Archimedean idealization to liquids.

The work of Archimedes was not the only Ancient analysis of the lever available as a resource in the late sixteenth century. An influential treatment of it was included in the *Mechanical Problems,* attributed to Aristotle but probably written by one of his school rather than by Aristotle himself. The Aristotelian treatment of the lever involved considerations of displacements of a lever about its equilibrium position. If a lever is subject to a small displacement about its fulcrum then the two ends of the lever, representing the point of application of the effort at one end and the resistance of the load at the other, are displaced through circular arcs that have lengths proportional to their distance from the fulcrum. The system is in equilibrium when the ratio of effort to load is inversely proportional to the ratio of the speeds they acquire when displaced. This line of analysis when applied to the unequal armed balance yields the law governing the balance's equilibrium. As with the works of Archimedes, the *Mechanical Problems* became freely available in Latin translation in the sixteenth century.

Another significant tradition in mechanics that became available in the sixteenth century is represented in the writings of the thirteenth century figure Jordanus Nemorarius, who had access to the Greek tradition through a number of Arabic sources.[4] The treatment of the balance by Jordanus echoes that in the *Mechanical Problems* insofar as it involves analogous dynamical considerations, and resembles the works of Archimedes insofar as it is mathematically formulated. The account of the balance by Jordanus related the ratios of balancing weights to ratios of the corresponding displacements rather than to speeds of displacement and differed from the Aristotelian tradition in that respect. It also differed from that tradition through the extent to which the science of weight was cast in a mathematical form characteristic of Archimedes rather than Aristotelians.

The mathematical formulation of the science of weight was transmitted through the sixteenth century via figures such as Nicolo Tartaglia and Giovanni Battista Benedetti to the extent that it had become the norm that was taken for granted by Stevin and his contemporaries such as Guidobaldo del Monte and the young Galileo. All these sixteenth- or early seventeenth-century scholars adopted approaches in their science that can be described as Euclidean or Archimedean insofar as they presented their theories as a body of propositions derived from postulates that could be cast into a geometrical form and were sufficiently unproblematic to be granted at the outset.

There are very few references to sources or predecessors in Stevin's works on mechanics. He does mention the treatment of the balance which considers the cause of equilibrium 'to reside in the circles described by the extremities of the arms as appears in Aristotle's *In Mechanics* and his successors' only to reject it.[5] But there

---

[4] For overviews of the mechanics of Jordanus and its sources see Drake and Drabkin (1969, pp. 10–12) and Renn and Damerow (2012, pp. 27–31).

[5] Stevin makes this remark in an *Appendix to the Art of Weighing,* Dijksterhuis (1955, pp. 507–509).

is no room for doubt that as far as his *Art of Weighing* and *Elements of Hydrostatics* are concerned his main source and inspiration was the work of Archimedes, especially the latter's *On the Equilibrium of Planes* and *On Floating Bodies.* This much is manifestly apparent from the content and style of Stevin's two treatises and is made quite explicit by him in a note *To the Reader* at the beginning of his *Elements of Hydrostatics.*

> What was the cause that moved Archimedes to write that which he left to us in the Book of the things suspended in water, where he began to hit off Nature wonderfully, I do not know. But I do know and gladly confess that he was the cause which induced me to cast this matter into the form we have given it (Dijksterhuis 1955 p. 385).

Before I proceed to an analysis of Stevin's hydrostatics in the next chapter I comment on the style and content of Archimedes' *On Floating Bodies*, Stevin's starting point.

2.3 Archimedes on Floatation

The work *On Floating Bodies* is justly famous for its introduction of 'Archimedes' Principle' specifying the conditions under which a body floats in water or suffers an apparent loss of weight on immersion. A close look at Archimedes' derivation and deployment of it reveals that its focus was very much on floating and, what is more, floating under some rather special and artificial conditions. Archimedes initiated the attempt to understand floating, to be sure, but not in a way that offered much potential for developing a more general theory of hydrostatics.

On Floating Bodies was represented as a body of propositions derived from postulates that were to be granted, and bodies were represented as simple mathematical figures. Book I, in which floating and immersion is characterized in terms of what we now call Archimedes' principle, is quite short and involves just two postulates.[6] The first of them is introduced with no fanfare as the very first sentence of the book. It reads as follows:

> Let it be granted that the fluid is of such a nature that of the parts of it which are at the same level and adjacent to one another that which is pressed the less is pushed away by that which is pressed the more, and that each of its parts is pressed by the fluid which is vertically above it, if the fluid is not shut up in anything and is not compressed by anything else.[7]

Here Archimedes has attempted to characterize fluids in a way that will give him what he needs for his theory of floatation. We need to understand the meaning and

[6] The second postulate, that specifies that the upward thrust on a partially or fully immersed body acts vertically through its center of gravity, is exploited in the much longer Book II which is largely concerned with the stability of floating bodies and which lies beyond our present concerns.

[7] Dijksterhuis (1956, p.373). I have used the translation of Dijksterhuis rather than the unsatisfactory one in *The Works of Archimedes*, Heath (1950, p. 253). This latter work was originally published by Cambridge University Press in 1897.

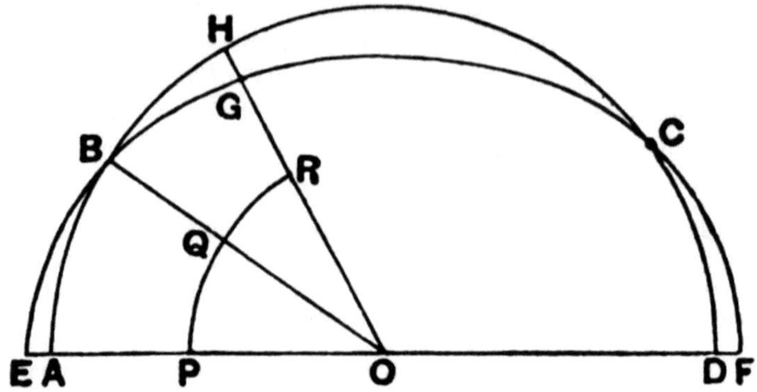

Fig. 2.1 Archimedes' spherical world of water (Heath 1950, p. 254)

content of this proposition and on what grounds it can be considered sufficiently unproblematic to warrant Archimedes' request that it be granted. To do so we need to attend to how Archimedes deployed Postulate 1 in his subsequent discussion.

After deriving a proposition related to a property of spheres which need not detain us, Archimedes proceeds to derive Proposition 2 which asserts that the surface of a fluid in equilibrium is the surface of a sphere centered on the Earth, depicted by O in Archimedes' diagram, reproduced as Fig. 2.1. It is clear that Archimedes has set up the problem in a way that involves a major idealization. To aid his intent to grasp the specific nature and behavior of fluids, he treats the presence of solids as an unnecessary complication, and removes them from the picture completely in his opening move. This might appear as an unrealizable goal insofar as liquids require a container to confine them. By considering a world consisting entirely of a sphere of water gravitating towards a center, marked O in the diagram, Archimedes has chosen a situation in which water confines itself. The final clause in the statement of Postulate 1 is satisfied by removing solids altogether and by adding the further idealization that the liquid is to be subject only to the pushes arising from its own weight.

Proposition 2 requires that in equilibrium the surface of the water represented in Fig. 2.1 will be the hemisphere represented in two dimensions by the semi-circle ABHCD, as opposed to some other surface such as EBGCF. To prove this result, Archimedes considers the weighing down of the liquid on two equal and neighboring planes represented by equal arcs PQ and QR. Were the surface to be EBGCF then the weight bearing down on PQ would be the body of liquid represented by EBQP, and that bearing down on QR would be the body of liquid represented by QBGR. But EBQP is greater than ABQP while QBGR is less than QBHR (which is equal to ABQP). Since the neighboring planes are pressed unequally, then, by Postulate 1, 'that which is pressed the less is pushed away by that which is pressed the more' and equilibrium will not obtain. Only when the surface is part of a sphere will PQ and QR be pressed equally and equilibrium result. That is what Archimedes had set out to prove.

It is instructive to pinpoint the precise way in which an understanding of the nature of fluids as such figures in Archimedes' argument. Firstly, it is involved in his assumption, included in Postulate 1, that 'each of [the fluid's] parts is pressed by the fluid which is vertically above it'. Each part of fluid QBHR, say, weighs towards the center, O, and hence so does the combined weight. The fluid outside of QBHR can be ignored because, being fluid, it is not attached to it in any way.[8] The idea that neighboring parts of a fluid do not constrain each other from moving because they do not cohere can also be taken to lie behind the idea that if part of a fluid in pressed more than a neighboring part then it will move. To some degree, then, the assumptions lying behind Postulate 1 have a basis in a common distinction between solids and liquids that was centuries old. Solids cohere in a way that liquids do not, which is why the latter flow and spill, why a hand can be moved slowly through them and so on. However, as will shortly be discussed, such common sense knowledge needed to be refined, organized, and eventually transcended, in a way that Archimedes failed to appreciate, even if it is only floatation rather than hydrostatics as a whole that is in question. Before taking stock of Archimedes' theory in this broader perspective, I summarize how it was developed by him beyond Proposition 2 to explain floatation.

Having eliminated solids from the picture to establish the equilibrium conditions for his sphere of water centered on the earth, Archimedes reintroduced them, but in a circumspect way. He imagined portions of cones of water, such as QBHR, to be replaced by a solid of the same shape and size. First the solid is presumed to be of equal density to that of the liquid. Archimedes maintains that in such a case the solid will weigh down on the plane, represented by QR in our example, to the same degree as the liquid it has replaced did, and so will have no effect on the equilibrium. The introduced solid will float with its upper surface coincident with the liquid surface. If the introduced portion of a solid cone is less dense than the liquid, then the liquid at QR will be pressed less than the liquid at PQ. Consequently, the solid will be 'pushed away' according to Postulate 1. Equilibrium will be restored once QR and PQ are pressed to the same degree and that will be the case once the solid has risen to such a degree that its weight, pressing on PQ, is the same as the water pressing on QP. This gives Archimedes his Proposition 5, 'Any solid lighter than a fluid will, if placed in a fluid, be so far immersed that the weight of the solid will be equal to the weight of the fluid displaced'. A similar line of argument was applied to immersed solids more dense than the liquid, yielding Proposition 7, 'a solid heavier than a fluid will, if placed in it, descend to the bottom of the fluid, and the solid will, when weighed in the fluid, be lighter than its true weight by the weight of the fluid displaced'.

[8] This assumption does not hold in the case of a solid because its parts cohere. A vase with a narrow base does not press on the floor with a weight equal solely to that of the vase plus contents that lies vertically above the base. As a matter of fact Archimedes' assumption does not apply to liquids either once we move beyond the special circumstances involving a spherical distribution of water. The degree to which water presses on the base of a vertical pipe with a right-angled bend in the middle is not equal to the weight of the water that lies vertically above the base.

So here we have theorems about floating of the kind we learn about at school and attribute to Archimedes. However, Archimedes was not merely intent on identifying his theorems. He presented himself as having rigorously derived them from grantable assumptions. In this latter respect, his arguments left quite a bit to be desired. Before pursuing that issue, I point out that there is a way of interpreting Archimedes' account of floating that lends plausibility to it and which was adopted and developed by later scholars, as we will see.

Here is an account of floatation that is not the one explicitly articulated by Archimedes, but is close to it and can, not unreasonably, be seen as representing the essence of it. It is a line of reasoning that straightforwardly yields Archimedes' Propositions 5 and 7. When solids are immersed in water, whether the water is in an enclosed vessel or in the open sea, it displaces an amount of water that, as a consequence, rises because there is nowhere else for it to go. Floating occurs when the weighing down of the displaced liquid exactly balances the weight of the solid. Floating is understood by analogy with the balance. In the latter, the weights on each of the balance arm press downward. However, because the weights are connected by a rigid balance arm, one weight, in pressing down, presses up on the other. Balance occurs when each weight is countered by the upward thrust communicated to it from the other via the balance arm. In the case of floating, both an immersed solid and the liquid displaced weigh down. The downward pressing of one results in an upward thrust on the other by way of the fluid that connects them. Equilibrium, that is, floating, occurs when the upwards thrust on the solid is equal to its weight.

Let me now return to the argument as Archimedes presented it and point to some limitations. As we have seen, his arguments are applied to situations that are subject to some rather severe simplifying assumptions. First we have a sphere of water with the same center as the center of the earth. Next we have situations where some solids with rather special shapes are introduced into the picture. But this is where it stops. Archimedes did not go beyond this to show how his theory could apply to more realistic situations, such as solids of a variety of shapes floating in variously configured vessels or ships floating in a dock. Further, his talk of pressing on neighboring parts, as it appears in Postulate 1 and is deployed thereafter, begs some important questions.

In a wide range of circumstances, it is not difficult to see how Archimedes' account could be extended from the ideal circumstances he discusses to floating in vessels. If a solid is immersed in a regular vessel then Archimedes' line of argument can be pursued by invoking neighboring planes at the same level and considering the weight bearing down on them. But there are arrangements that would have confounded Archimedes. Suppose water is contained in a vessel that has the shape of an inverted T. Archimedes assumption that the water weighs down on a plane by virtue of the weight of water that lies vertically above it yields the wrong answer for the force on that part of the base that is not below the vertical arm.

E. J. Dijksterhuis saw a problem with the extent to which Archimedes' discussion is confined to neighboring parts of a liquid.

> On this spherical surface [one within the body of a liquid and equidistant from the center] he [Archimedes] takes two adjacent equal surfaces *a* and *b*, and he now seems to imagine that if the pressure on *a* is greater than the pressure on *b*, the fluid in the sector having *a* for its section will force the fluid on which *b* lies upwards via the center of the universe. We do not hear what happens to the fluid in other sectors, whether this is assumed to be motionless or whether it is also displaced; the postulate is explicitly confined to adjacent surfaces at the same level. (Dijksterhuis 1956, p. 378)

It is true that Archimedes focuses his discussion on adjacent parts of a liquid at the same level and that his talk of the part being pressed the more being 'pushed away' is unspecific about this pressing and the details of the ensuing motion. Nevertheless, I believe Dijksterhius' criticism to be misplaced. It is offset if we understand the focus to be on equilibrium states. If any neighboring parts of a liquid at the same level are pressed differently they will not be in equilibrium because their parts do not cohere in a way that they do in solids. The part that is pressed the more will be pushed away. Exactly how it is pushed away is not relevant. What is relevant is the state that will restore equilibrium. Archimedes' discussion of the replacement of a body of liquid by a body of solid compares two equilibrium states, the one where the liquid is in equilibrium with itself and a second where part of it is replaced by a solid resulting in a new equilibrium state. There is no need for Archimedes to describe the motions or pressures involved in the transition from one to the other. There is one detail that Archimedes could add. According to his theory the change in equilibrium brought about by the introduction of a solid into a liquid involves one spherical surface being replaced by another that is slightly higher.

If we are to understand floating by identifying the pushes responsible for it, then the focus on equilibrium states introduced above leaves something to be desired. Archimedes' theory has the implication that the weight of liquid displaced by a solid, which acts downward, somehow results in a force on the solid that acts upwards. This conversion of a downward force into an upward one implies that the forces involved in the transition somehow does a 180° turn, but no account of how this comes about is given. The same point applies to the interpretation of Archimedes' principle that employs an analogy with the balance. In the latter the change from a downward to an upward force is due to the rigidity of the balance arm. What, exactly, is the corresponding account of a mechanical linkage in the case of floating in liquids?

As I have stressed, Archimedes was intent on providing an account of floatation and made no claims to be doing more than that. However, if we do consider the extent to which it could serve as a starting point for a theory of hydrostatic phenomena in general, then it is possible to identify shortcomings. They stem from the nature of the idealizing assumptions that Archimedes made and never relaxed and from the exclusive focus on forces acting in a vertical direction.

In Postulate 1 that serves as the basis of a large part of his theory Archimedes specified that the liquids he was to treat of are 'not shut up in anything'. Containing vessels do not enter the scene, and solids enter the picture only as regularly shaped objects that replace an equivalent volume of liquid. Since this assumption is never qualified or relaxed, Archimedes' theory has no capacity for dealing with a host of situations involving liquids at rest in variously configured vessels, and in which his assumption that liquids bear down on a plane by virtue of the weight of liquid that lies vertically above it is false. When Roman engineers were concerned with manufacturing pipes strong enough to bear the pressure of the water from the lake that was its source, they were well aware that those pressures depended on the height of the lake above the pipe, it being obvious that the lake was not vertically above the pipe. By the time Stevin turned his attention to hydrostatics he was able to stress the point in a more artificial case, the case represented in Fig. 1.1 in the previous chapter, where a stopper is borne down by a force greatly in excess of the weight of liquid that lies vertically above it.

A related, and more far-reaching, problem is the extent to which Archimedes' account of floatation is confined to specification of vertically acting forces only, the downward action of weight and the upward thrust of liquids on solids wholly or partially immersed in them.[9] Archimedes' account of flotation offers little scope for extension to non-vertical pressures such as those responsible for the flow of wine from a faucet in the side of a wine-barrel or the pressure exerted by water against a lock gate.

Stevin explicitly referred to the work of Archimedes as providing him with his starting point in hydrostatics. He did attempt to deal with forces exerted by liquids in directions other than the vertical. Stevin did include the horizontal force against lock gates in his deliberations. His ingenious efforts in this respect are described and evaluated in the next chapter. We will see that Stevin's theory of 1586 represents only the beginning of our story. The end lay more than a century away.

References

Clagett, M. 1959. *The science of mechanics in the middle ages*. Madison: University of Wisconsin Press.

Dijksterhuis, E.J. 1955. *The principal works of Simon Stevin, Volume 1, Mechanics*. Amsterdam: Swets and Zeitlinger.

———. 1956. *Archimedes*. New York: Humanities Press.

Drake, S., and I. Drabkin. 1969. *Mechanics in sixteenth-century Italy: Selections from Tartaglia, Benedetti, Guido Ubaldo and Galileo*. Madison: University of Wisconsin Press.

Duhem, P. 1991. *The origins of statics: The sources of physical theory*. Dordrecht: Kluwer Academic Publishers.

Heath, T.L. 1950. *The works of Archimedes*. New York: Dover.

Moody, E.A., and M. Clagget. 1960. *The medieval science of weights*. Madison: University of Wisconsin Press.

[9] When Archimedes, towards the end of Book I, introduced his second postulate he was already expressing himself in terms of upward thrusts as counters to weight and a large part of Book II considers the stability of floating bodies of various configurations in terms of the vertical line of action of the weights and the upward thrusts that oppose them. Postulate 2 reads 'Let it be granted that bodies which are forced upwards in a fluid are forced upwards along the perpendicular which passes through the center of gravity' (Heath 1950, p. 261).

Palmieri, P. 2008. Breaking the circle: The emergence of Archimedean mechanics in the late Renaissance. *Archive for History of the Exact Sciences*. 62: 301–346.
Renn, J., and P. Damerow. 2012. *The equilibrium controversy: Guidobaldo del Monte's critical notes on the mechanics of Jordanus and Benedetti and their historical and conceptual background*, Max Planck Research Library for the history and development of knowledge. Sources 2. Berlin: Edition Open Access. http://www.edition-open-access.de.
Rouse, H., and S. Ince. 1963. *History of hydraulics*. New York: Dover.
Van Leeuven, J. 2016. *The Aristotelian mechanics: Texts and diagrams*. Dordrecht: Springer.

Chapter 3
Beyond Archimedes: Stevin's *Elements of Hydrostatics*

Abstract The content of Stevin's hydrostatics was original and became a cornerstone of the advances in hydrostatics that were to follow it. Stevin grasped the fact that the force on a solid surface in contact with water is independent of the orientation of that surface and depends only on its depth beneath the uppermost surface of the water. Some of Stevin's proofs were highly ingenious. However, his postulates, the most significant of which acknowledged that the depth to which a vessel sinks is proportional to the weight it carries, were insufficient to yield the content of his theory as deductive consequences. In a way that he did not make explicit, Stevin inserted into his theory features of the behavior of water with which he was familiar as a hydraulic engineer but which were not licensed by his postulates. For instance, he assumed, rather than proved or explained, that water presses horizontally against a vertical surface, a phenomenon with which Stevin was familiar through his dealings with lock gates. It was only during the course of the seventeenth century that the facts about hydrostatics identified by Stevin were explained and adequately theorized. Those developments required the recognition that experiment needs to be read as supplying evidence for hydrostatics rather than as being mere applications of it, as Stevin had assumed.

3.1 Simon Stevin: Scientist/Engineer and Practical Mathematician

Simon Stevin (1548–1620) was characteristic of a new kind of mathematically trained engineer who came to prominence in the sixteenth century. Those scholars, including the likes of Nicolo Tartaglia (1500–1557) and Giovanni Battista Benedetti (1530–1590), were well versed in the knowledge of practitioners such as navigators, architects, surveyors and mechanics. They reflected on the practitioners' knowledge and attempted to theorize it, taking advantage of their mathematical skills. These scientist/engineers or practical mathematicians typically worked outside of University contexts, often supported by wealthy patrons. I cannot improve on the characterization of this group by Matthias Schemmel (2008, p. 15).

> This group had formed in the course of the late fifteenth and sixteenth centuries, first on the European continent and, in the latter half of the sixteenth century, also in England. The

A.F. Chalmers, *One Hundred Years of Pressure*, Archimedes 51,
DOI 10.1007/978-3-319-56529-3_3

> practical mathematicians attempted to solve practical problems in navigation, surveying, shipbuilding, fortification, gunnery, and similar fields of contemporary practical concern. They handled engineering problems on commission, held lectures on practical issues, instructed seamen, and designed new instruments. In view of this they may be considered practitioners. They were, however, distinct from the majority of practitioners by their reflection on the practical knowledge. They strived for a mathematization of that knowledge, they edited ancient works on mathematics and translated them into the vernacular to make them accessible to a wider readership, they systematically scanned the existing practical knowledge for is usefulness, they performed 'experiments', the results of which they hoped to be of direct practical use, they wrote manuals and books to communicate their knowledge and emphasized the importance of their knowledge to society in programmatic writings.

Prior to his study at the University of Leyden (1581–1583) Stevin had worked in financial administration in his hometown, Bruges, and as a cashier and bookkeeper in Antwerp. By the time he matriculated at the University of Leyden in 1583 he had already published books on *Tables of Interest* and on *Geometrical Problems*. By that time Stevin was also involved in technical innovations, especially associated with draining and dredging, applying for patents for several of his inventions. In 1588 he formed a partnership with his friend Johan Cornets de Groot in order to put his inventions to practical use. From the mid-1590s Stevin was in the service of Prince Maurice of Nassau who was commander-in-chief of the States' Army. He acted as Maurice's tutor in practical mathematics and several of his subsequent publications, on fortifications, optics, astronomy, tides and so on, took the form of textbooks designed to educate his patron and employer. These later works involved an opportunist and far from uniform combination of empirical and mathematical considerations which contrasts with the strict mathematical character of his earlier works. In 1608 Stevin published a compilation of his works, his *Mathematical Memoirs*. A Latin translation of them was included as a supplement, extending his readership beyond those fluent in the Dutch language.[1]

The two books by Stevin that are relevant for this study are *The Art of Weighing* and *The Elements of Hydrostatics*, written soon after his graduation, before his partnership with de Groot commenced, but at a time when he was already involved in practical hydraulic projects.[2] The works were presented as mathematical ones in the style of, and heavily influenced by, Archimedes. They were published in 1586, more than half a decade before Stevin's association with Prince Maurice began.

[1] For more detail on Stevin's life see Dijksterhuis (1970).

[2] Original versions of those two works in Dutch are reproduced, together with an English translation in Dijksterhuis (1955), pp. 35–295 for *The Art of Weighing* and pp. 375–483 for *The Elements of Hydrostatics*. To each work is appended a section dealing with practical applications, *The Practice of Weighing* (pp. 297–373) and *Preamble of the Practice of Hydrostatics* (pp. 485–501). Also included in this source is Stevin's *Appendix to the Art of Weighing* (pp. 503–521) and *Supplement to the Art of Weighing* (pp. 523–607).

3.2 Stevin's *The Art of Weighing* and the Euclidean Ideal

The Art of Weighing is presented as a body of theorems derived from postulates. The subsequent theory, outlined in Book 1 of the work, extends treatment of vertically acting weights to non-vertical actions mediated by inclined planes and pulleys. Book 2 is concerned with the calculation of centers of gravity of a range of plane and solid figures. The mathematical techniques employed include splitting the action of a weight into its components, adding weights using the parallelogram of forces, and locating centers of gravity by appealing to limits which usage marked a step towards the infinitesimal calculus.

Stevin followed his treatise on *The Art of Weighing* with a sequel on *The Practice of Weighing*. This included practical techniques for the determination of centers of gravity and of measuring weights. He also described various machines employing ingenious arrangements of levers and pulleys, including an exceedingly powerful machine that he called 'The Almighty', involving a wheel and axle driven by a crank handle via a system of gear wheels. Insofar as this machine was capable of hauling heavy ships over sand, it extended the notion of a machine beyond the ability to raise weights to that of overcoming other resisting forces.

Most important for our concerns in this book is not so much the content of Stevin's *Art of Weighing* as the style of reasoning that was represented by it. This style, which I refer to as the Euclidean ideal, had become the norm as far as sixteenth-century sciences inspired by Archimedes are concerned. I introduce the notion, not as a philosophical category, but simply as a convenient term to describe four features possessed by the style of reasoning involved. First, theories are to be presented as deductive consequences of postulates that are sufficiently unproblematic to be granted at the outset. Second, theories are abstractions insofar as they focus on certain aspects of physical situations while neglecting others, exemplified by the way in which geometry considers shapes and dimensions of systems while ignoring their material constitution. Third, the theories are applied to idealized sets of circumstances as distinct from more complicated real-world set-ups. For example, Archimedes calculates centers of gravity of ideal planes such as triangles as opposed to wheelbarrows filled with sand. Fourth, because propositions of a science are established by deriving them from postulates that are granted, application of the propositions to real-world situations is not to be construed as supplying evidence for those propositions. To imagine that a mechanical theory developed according to the Euclidean ideal can be supported by experiments with balances, say, is as wrong-headed as to suppose that the theorems of geometry are to be confirmed by measurements on material systems.

Stevin was quite explicit about the extent to which his *Art of Weighing* conformed to what I have called the Euclidean ideal. First, we find him making the point that the propositions of his *Art of Weighing* are to be deduced from postulates granted at the outset.

> Since some matters of an elementary nature are common knowledge, and need not be proved, while other matters of a more veiled character might give the critic cause to criticize

> that which does not deserve criticism, we shall, after the matter of mathematicians, before arriving at the propositions, postulate that the following things be granted (Dijksterhuis 1955, p. 111)

Second, we have mechanics involving abstractions similar to those involved in geometry.

> Whereas geometry relates to the magnitudes of figures, not their gravities, holding only those to be equal or unequal whose magnitudes are equal or unequal, the art of weighing on the contrary relates to their weights, not their magnitudes, holding those to be equal whose weights are equal or unequal. And just as the chief task of the former consists in examining the ratios, proportions and properties of their magnitudes, so the task of the latter consists in examining the ratios, proportions and properties of their gravities or weights, the description of which is the object of this treatise. (Dijksterhuis 1955, p. 97).

Third, idealization is manifest by the preponderance of geometrical diagrams accompanying the proofs of the propositions of Stevin's statics together with the absence of depictions of real experimental set-ups. Stevin explicitly addressed the idealization involved in his assumption that neighboring verticals are parallel, defending it on pragmatic grounds since 'the difference [between parallel verticals and verticals directed towards the center of the earth] is imperceptible in all things weighed by us' (Dijksterhuis 1955, p. 115).

As far as the application of the propositions of the *Art of Weighing* is concerned, this is treated by Stevin in a separate volume, *The Practice of Weighing,* with no hint of the notion that the considerations of the second volume are providing evidence for the propositions derived in the first.

The Euclidean ideal as I define and employ it is distinct from the notion of 'mixed mathematics' used to describe physical theories that employed mathematics, including hydrostatics. Mixed mathematics was a category used by scholastics of the time in their attempt to accommodate sciences such as hydrostatics to the philosophy of Aristotle. According to Aristotle, physical theories, or natural philosophy, and mathematics both have their legitimacy but are distinct. The spherical shapes and circular motions of planets is dealt with by the former, whereas the latter is concerned with the properties of spheres and circles as abstracted from material objects. A science such as hydrostatics, concerned with the physical behavior of liquids but cast in mathematical form, is 'mixed' in the sense that it involves an opportunist conjunction of two quite different forms of knowledge. Development of the notion of mixed mathematics was on ongoing philosophical enterprise in the sixteenth and seventeenth centuries and is one worthy of scholarly investigation.[3] However, it is not one to which I aim to contribute in this book. My focus on the Euclidean ideal, as opposed to mixed mathematics, enables me to highlight key changes in subsequent versions of hydrostatics insofar as they departed from that ideal.

---

[3] Peter Dear (1995) has studied in detail scholastic attempts, especially by Jesuit scholars, to accommodate the developing mathematical sciences into their philosophy during the Scientific Revolution, noting in particular the modifications that were made in the scholastic understanding of 'mixed mathematics'. Stephen Gaukroger and John Schuster (2002, pp. 335–338) discuss similar issues in the context of Descartes' hydrostatics.

3.3 Hydrostatics, the Euclidean Ideal and Stevin's Postulates

In the dedication of his *Elements of Hydrostatics* to the burgomasters and rulers of Nuremburg Stevin remarked on the importance in the Low Countries of practical dealings with water. He portrayed his *Elements* as important in that context insofar as that work was to identify 'the causes of the statical properties of water' thereby revealing 'effects not seen or known by our predecessors' (Dijksterhuis 1955, p. 381–3). In the ensuing note *To the Reader* Stevin made it explicit that he saw his efforts as inspired by and modeled on the work of Archimedes. The extent to which the theory was to employ abstractions was made clear by his elaboration of the point.

Just as Euclidean geometry involves notions of point, straight line and plane so hydrostatics requires notions peculiar to it. Here Stevin introduced what my contacts at the Max Planck Institute for the History of Science would call a new 'mental object'.[4]

> It is to be noted that as in the *Elements of Geometry* the point is to be taken without length, the line without breadth, the plane without thickness, in the same way it is necessary in the *Elements of Hydrostatics,* in order to deal therewith geometrically, to assume vessels without any corporeal magnitude and without any weight. These we have called, in accordance with their properties (for new arts call for new words) surface vessels (*vlackvat*), since their material consists of surfaces, as will be said in the 7th definition hereafter. (Dijksterhuis 1955, p. 385)

The 7th definition referred to reads 'Surface vessel is the complete geometrical surface of a body, conceived as separable therefrom'. This definition does not convey quite what Stevin has said in his note to the reader and which is clear from his subsequent usage of the notion of surface vessel. These vessels abstract from weight and thickness but are not merely geometrical surfaces. They are also solid or corporeal insofar as they are capable of supporting or containing liquid.

A related technical 'mental object' is Stevin's notion of a 'bottom' *(bodem)* defined as 'any plane against which rests any water' (Definition 8). These also go beyond being merely geometrical insofar as they prevent the passage of water. Such planes include vertical and slanting sides of a container, not just horizontal ones at the base of containers as is suggested by connotations of the word 'bottom'.

Stevin talked of his introduction of new terms as if it were merely a matter of defining words. He attributed his success in this regard to the merits of the Dutch language.[5] However, new 'arts' require not simply new words but new concepts.

---

[4]According to Jürgen Renn (in Schemmel 2008, p. viii) these are 'knowledge representation structures' which can have longevity as resources to be exploited by scientists and survive changes from one theory or conceptual framework to another. Examples include the notion of a center of gravity or a model of a balance that abstracts from its physical features other than weight and the presumed rigidity of the balance arm. I would prefer the term 'theoretical model' rather than 'mental model' to avoid the impression that these models exist only in the mind. Stevin's 'surface vessels' are capable of holding water!

[5]Stevin introduced his *Art of Weighing* with a *Discourse on the Worth of the Dutch Language* (Dijksterhuis 1955, pp. 59–65).

The definition of a straight line in geometry is adequate to the extent that it captures the essential features of idealized versions of such things as stretched strings and lines of sight. In like manner, Stevin's definition of a surface vessel is adequate to the extent that it serves to capture or articulate some essential behavior of solids and fluids. As we shall see, and as was made explicit by Newton, the task of putting hydrostatics on a secure theoretical foundation hinged on the formulation of an adequate definition of fluids that would serve to distinguish them from solids. Once this is appreciated, the distinction between definitions and postulates presupposed in Stevin's presentation becomes blurred.[6] 'Surface vessels' and 'bottoms' acquire a role and meaning not simply through being defined but by being incorporated into postulates and aiding the derivation of propositions related to the way in which fluids behave. We will soon be investigating the extent to which they could fulfill such a role for Stevin.

Stevin did not explicitly acknowledge the task of grasping theoretically the distinction between solids and liquids as an important aspect of his problem. He did not define, or otherwise address, the way in which liquids are essentially different from solids. In his Definition 6 he did define 'solid body' as 'one whose matter does not flow, and through which penetrates neither water nor air'. Perhaps this can be taken as implying that fluids, as distinct from solids, are characterized by their capacity to flow. Here Stevin was doing little other than trade on what is implied in everyday knowledge and language.

Our discussion of the abstractions introduced and discussed by Stevin early in *The Elements of Hydrostatics* has led us to begin to consider the content that Stevin introduced into his hydrostatics by deploying them. Since it was Stevin's purpose to derive the propositions of his hydrostatics from postulates to be granted at the outset, it is pertinent to raise the question of how the content of his theory entered into those postulates. Once this is done then a problem can be raised which other historians seem to have missed.[7] Stevin's postulates are insufficient to yield the important content of his hydrostatics. I will argue that Stevin violated the Euclidean ideal by inserting into his theory knowledge not covered by his postulates and which presumably had its roots in his experience as a hydraulic engineer. I do this, not because

[6]When Stevin identified definitions in addition to postulates as forming the basis from which he would derive the propositions constituting his theory he was following the example of Euclid rather than Archimedes. The latter listed only postulates, taking for granted the kinds of matters Euclid and Stevin made explicit with their definitions.

[7]Pierre Duhem (1905, p. 603), one of the few historians to have paid detailed attention to Stevin's hydrostatics, may well have been justified in acclaiming the latter for its novelty, but, as we shall see, his extension of that acclaim to include rigor cannot be allowed. Dijksterhuis (1955, p. 377), another historian to have given serious attention to Stevin's hydrostatics, also failed to pick up on all of the shortcomings that I will identify. The attribution of rigor to the deductions in Stevin's hydrostatics seems to have remained unchallenged in more recent literature. For example, in their otherwise excellent historical analysis of early work of Descartes on hydrostatics, Gaukoger and Schuster (2002, pp. 539–540) impute to him a firm belief in Stevin's rigor and seem to endorse this judgment themselves, even though doing so is not material to their historical case.

I wish to belittle Stevin's insights, but rather to identify in an illuminating way the work that remained to be done by his successors.

It is not over-generous to summarize the content of Stevin's hydrostatic theorems as follows:

> Normal to any element of solid surface bounded by a liquid, whatever its orientation, there is a force equal to the weight of a column of liquid with cross-section equal to that of the surface element in question and a height equal to the depth of that surface element beneath the liquid surface.[8]

The content of this assertion is sometimes referred to as Stevin's law and I will follow that example. Stevin's law is true from a modern point of view and, moreover, gives hydrostatics much of what it needs as far as effects arising from the weight of the liquids themselves are concerned. A perusal of Stevin's postulates reveals that they are insufficient to yield Stevin's law.

The only significant candidate in this respect is Stevin's Postulate III. It reads as follows:

> The weight causing a vessel to sink less deep to be lighter, but the weight causing it to sink deeper to be heavier, and that causing it to sink to the same depth, equally heavy (Dijksterhuis 1955, p. 397).

I do not suggest there are grounds for having qualms about granting Stevin this postulate. It can be taken as representing common knowledge of his day. What I do wish to point out is that it does not contain enough to form the basis of his hydrostatics. In particular, insofar as it refers only to the downward action of weight and vertical resistance to it, Postulate III cannot yield the isotropic hydrostatic forces involved in Stevin's hydrostatics and implied by Stevin's law. The thrust of my critique of Stevin's theory will be that in deriving his theorems he introduced or implied key assumptions that are neither included in, nor are consequences of, his postulates. This was a matter that needed to be addressed by later workers who attempted to build on Stevin's pioneering efforts.

3.4 Stevin's Derivation of Archimedes' Principle

The first proposition of Stevin's hydrostatics asserts that a volume of water surrounded by water will retain any place given to it. His argument is that, were such a volume to move, the overall situation that eventuates would be identical to the one at the outset, so the volume would move again, and so on. Stevin rules out this possibility because it leads to perpetual motion, a mode of argument that he had

[8] My use of the term 'surface element' serves to make explicit in a modern way what is only implicit in Stevin's own words. The notion of a surface element is necessary because the extent to which a liquid presses on a solid surface is a continuous function of the depth of that solid surface beneath the uppermost surface of the liquid. This much is implicit in Stevin's hydrostatics and is handled by him through the use of limits, an example of which we will shortly be discussing.

already employed in his celebrated treatment of the inclined plane in *The Art of Weighing*.[9] The precise content of the principle that asserts the impossibility of perpetual motion is unclear and its status problematic. Is the continuing motion of a pendulum in a vacuum an example of perpetual motion or not? To the extent that it is, it violates the principle in question.[10] Such an objection could have been put to Stevin. Other objections became possible in light of subsequent developments. Robert Boyle questioned Stevin's appeal to the impossibility of perpetual motion on the grounds that, to the extent that it is plausible that the heat of a body in thermal equilibrium consists in the motion of its component corpuscles, those motions must in some sense be perpetual (Boyle 1999, Vol. 5, p. 268). The uniform motion of a body free of forces asserted by Isaac Newton's first law of motion is an example of perpetual motion unless the latter is defined in a way that excludes it.

I do not wish to make much of this particular objection to Stevin's argument. I do, however, wish to raise the question of why Stevin saw it as necessary to *prove* the fact that a body of water will be in equilibrium in water. There were alternative courses open to him. One is that he could have taken the assertion as a brute fact to be granted on the basis of its obviousness. After all, a large volume of water does remain quiescent if allowed to settle. Another option, a version of which was taken by Newton in his hydrostatics as we shall see, is to take the symmetry of the situation to imply that there is no reason for a submerged volume of water to move in one direction rather than another. In any case, is it not the fact that the equilibrium of water in water is as unproblematic a circumstance as the one that Stevin was prepared to grant, in his Postulate 3, specifically that a vessel sinks more or less deep depending on the size of the weight it carries?

I believe there is an answer to that latter question. It is one thing to accept that a body of water within water is in equilibrium. It is another thing to understand why that is the case. There is a distinction between a fact and a reasoned fact, as Aristotle had insisted. As we shall see, Boyle invoked this kind of point in the context of hydrostatics when he insisted that not only is it important to appreciate *that* various assertions involving the behavior of liquids are true, but also to appreciate *why* they are true. We can understand Stevin as wishing to deepen an understanding of hydrostatics through an understanding of why its various propositions are true.

There is a reason why Stevin may have been prepared to accept his Postulate 3 as a brute fact to be granted, without further explanation. He wrote at a time when an understanding of simple mechanical machines via the science of weight was well developed and commonly accepted. The main cause at work in the simple machines was weight. Weight as a cause was simply accepted as such and no need was expressed or attempt made to explain it at some deeper level by those who developed the science of weight. Such a stance was implicit in the works of Archimedes that were Stevin's starting point. We have examined, in Sect. 2.3, the extent to which Archimedes' account of floating was dominated by considerations of weight. Once we view Stevin's hydrostatics as an attempt to extend the science of weight beyond

[9] The relevant passage in *The Art of Weighing* is in Dijksterhuis (1955, p. 179).

[10] Objections along these lines are raised by Dijksterhuis (1955, p. 179, n.1 and 1970, p. 54).

the simple machines so as to accommodate liquids we can begin to see why he was prepared to include claims about weight in his postulates but not general claims about the behavior of water.

I have digressed somewhat from the point of this Section, namely, to follow Stevin's path to a derivation of Archimedes' principle. So let me return to it. Stevin considered a cubic surface vessel containing water and immersed in a larger container of water. According to Proposition 1, whose 'proof' we have discussed, the cube will rest in equilibrium. We next imagine the water in the cube to be poured out and the cube returned to its position. Postulate 3 ensures that the cube will now have a tendency to sink less deep. In Stevin's words it will have a 'levity' that is assumed by him to be equal to the weight of the absent cube of water. We now imagine the cube to be filled with a solid that is denser than water. Again according to Proposition 3, the levity of the cube when empty will now have superimposed on it the weight of the cube of solid. That solid cube will cause the containing cube to sink deeper than when it was full of water by an amount that is equal to the weight of the cube acting downward minus the levity of the empty cube acting upward, where that levity has already been assumed to be equal to the weight of water that fills the cube. The result is Stevin's Proposition 8, 'the gravity [i.e. weight] of any solid body is as much lighter than in air as is the gravity of the water having the same volume', one way of stating Archimedes' Principle.

Stevin's proof can be faulted on the grounds that it introduces a quantitative interpretation of Postulate 3 that goes beyond what the postulate asserts, for instance when it is assumed that the levity of the cube when empty is equal to the weight of the water required to fill it. But this point notwithstanding, it is fair to say that the loss of weight by an immersed solid and the phenomenon of floating which it serves to explain are better understood with the benefit of Stevin's deliberations than without it. However, the remark by E. J. Dijksterhuis (1970, p. 66) that 'the clearness of Stevin's insight into the phenomena exhibited by a solid immersed in a liquid leaves nothing to be desired' goes a little too far. Dijksterhuis (1970, p. 65) himself drew attention to the fact that Stevin did not explain Archimedes' principle by considering the forces acting on an immersed body, as is often implied in modern textbooks.

From a modern point of view one might wonder why Stevin introduced the surface vessel into his discussion of the equilibrium of water in water. Why not simply consider a cube of water within water? Had he done so, he would have come much closer to introducing a notion of pressure as acting across planes within the body of a liquid as Boyle and Newton were to do around a century later. Stevin needed his surface vessel in order to link his proof to Postulate III, which related the depth at which a vessel floats to the weight it carries. The fact that Stevin needed to cast his proof in such a form highlights the significance of the absence of any characterization of liquids and their behavior among his postulates.

Up to this stage my discussion has been confined to the first eight propositions of Stevin's hydrostatics that were concerned with the accommodation into it of phenomena well known for centuries and already theorized by Archimedes. It is time to investigate how Stevin pushed his hydrostatics to a point where it could accommodate phenomena, some of them puzzling and surprising, not appreciated by Archimedes.

3.5 The Force on a Horizontal Plane

Stevin could cope with a number of otherwise problematic hydrostatic phenomena through his realization that the force exerted by a liquid on a horizontal surface, whether it presses on it from above or below, depends only on the area of the surface and its depth below the liquid surface and is otherwise independent of the total amount of liquid. Stevin's Proposition X, which expresses this fact, is unproblematic in the case of water contained in a regular cylindrical vessel where it implies that the force on the base of the cylinder will be equal to the weight of the water in it. In other cases the predictions of the theorem are by no means self-evident and to some degree counter-intuitive. It was Stevin who drew attention to the possibility that a weight of several hundred pounds could be needed to support a few ounces of water, a puzzle I introduced in Sect. 1.2 and illustrated with Fig. 1.1, p. 2.

Stevin's Proposition X reads:

> On any bottom of the water being parallel to the horizon there rests a weight equal to the gravity of the water the volume of which is equal to that of the prism whose base is that bottom and whose height is equal to that of the prism whose base is that bottom and whose height is the vertical from the plane through the water's upper surface to the base.

Stevin's proof of his theorem utilizes a thought experiment that involves replacing volumes of liquid by equal volumes of a solid whose density is equal to that of the liquid. I will explain what I take to be the substance of Stevin's argument by reference to the situation depicted in Fig. 1.1. Imagine that we start with a regular cylinder of water on the base CD with height, h. We now suppose that all the water, other than the small amount contained in EFBA, is replaced by a solid equal in density to the water. Provided the introduced solid is freely floating in the water, it is reasonable to suppose that the move will have no effect on the force on CD. Indeed, this will be a consequence of Stevin's Postulate III. However, what Stevin needs to complete his argument and arrive at the situation depicted in Fig. 1.1 is to move beyond the situation where the introduced solid floats to one where it is secured so as the form a solid container for the water EFBA. Stevin needed to, and did, assume that this move will have no effect on the force on the base CD either. This step in the argument went unacknowledged by Stevin and, what is more, it is problematic insofar as it goes beyond what is entailed by his postulates.

The assumption, that replacing a volume of water by the same volume of a solid of equal density leaves the force on the base of the container unchanged, is persuasive so long as the solid is floating freely in the water. Once the solid is attached to the sides of the original container to form one of a different configuration the situation is significantly different. A solid attached to the side of a container, unlike a freely floating one, does not press down on a surface beneath it at all. What is at

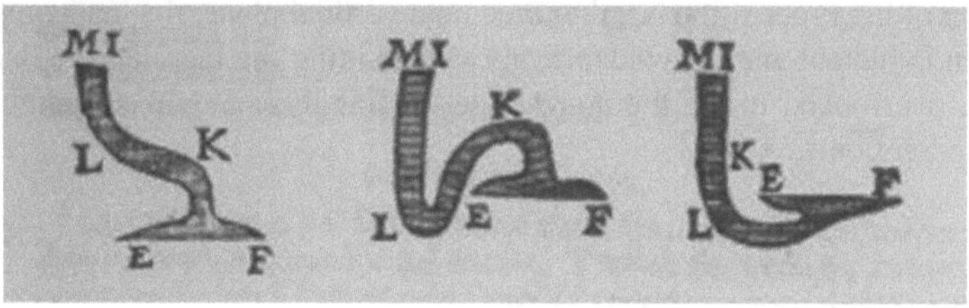

Fig. 3.1 Equal forces on the equal bases of variously contorted containers (Dijksterhuis 1955, p. 418)

issue here is a fundamental difference between the behavior of solids and liquids that Stevin has in effect assumed without acknowledgment or justification. If a solid rests, and weighs down, on a table the table acts on the solid with a force of reaction equal to the weight of the solid, resulting in equilibrium. If the weight is now supported by a clamp so that it no longer weighs down on the table the reaction of the table on the solid becomes zero. Suppose, by contrast, our solid weighs down on a liquid denser than it. Its weight will be countered by an upward thrust from the liquid resulting in equilibrium. Suppose that the floating solid is now clamped so that it no longer weighs down. In this case the reaction of the liquid does not fall to zero. The liquid will continue to press on the solid as before. Stevin's failure to distinguish between a body floating in water and a supported body resting against water results in an argument that lacks deductive rigor unless assumptions not included in or implied by Stevin's postulates are added.

Stevin extended his strategy to construct the irregular tubes of water depicted in Fig. 3.1. In each of the three cases shown the surfaces labeled EF are pressed equally, downwards in the first and second configurations and upwards in the third, provided the areas EF and the vertical height of MI above EF are equal.[11] Stevin arrived at these figures by starting with a rectangular prism of water with an upper surface including MI and imagining that the place of all water other than that displayed in the figures is taken by a solid equal in density to the water. Once again, the step from situations where the solid is freely floating in the water to those where the solid is fixed so as to form solid vessels capable of holding water is problematic and unjustified. The equality of the pressing on each of the horizontal surfaces represented by EF and its independence of the details of the contours of the tubes is an important and far from obvious fact that Stevin has pinpointed here. He identified the fact but he did not explain it by incorporating it into his theory in the way he portrayed himself as doing.

The precise nature of Stevin's argument here is not totally transparent and I am not totally confident that my interpretation of it does it justice. Fortunately, I do not need to rest my overall case on that interpretation. When it comes to Stevin's

[11] The height of MI above EF is more for the left hand configuration than in the two to its right in Stevin's figure as it is presented, but it is clear from the text that the depth of EF below the water surface should be the same in all three cases.

derivation of his Proposition XI, the force on a vertical plane, the inadequacy of his deduction is blatant and unavoidable, notwithstanding the fact that the inadequacy has been overlooked in the literature hitherto. I analyze Stevin's argument in the following Section.

3.6 The Force on a Vertical Plane

Stevin's Proposition XI specifies the force on a vertical plane. It reads as follows:

> Given a regular bottom whose highest point is in the water's surface: the weight resting against it is equal to half of the prism of water whose base is equal to that bottom, and whose height is the vertical from the highest point of the bottom to the plane parallel to the horizon through the lowest point of the bottom.

In the case depicted in Stevin's diagram, reproduced as Fig. 3.2, the force acting horizontally on the surface ACDE exerted by the rectangular prism of water AGFBDC is, according to Proposition XI, the weight of the wedge of water ACHDE which has surface ACDE as base and a height DH equal to CD, the depth of the lower edge of plane ACDE beneath the water surface. The proof of this is complicated by the fact that the degree to which the water presses on the vertical surface varies with height, from zero at the level AC to a maximum at the level ED. Stevin was able to meet the challenge posed by this complication.

Stevin calculates the force on the vertical plane by trapping it between upper and lower limits that he can show converge on each other, a technique that Galileo and Descartes were to use in the context of free fall several decades later. Stevin begins by dividing the prism of water into four slabs of equal height by equally spaced

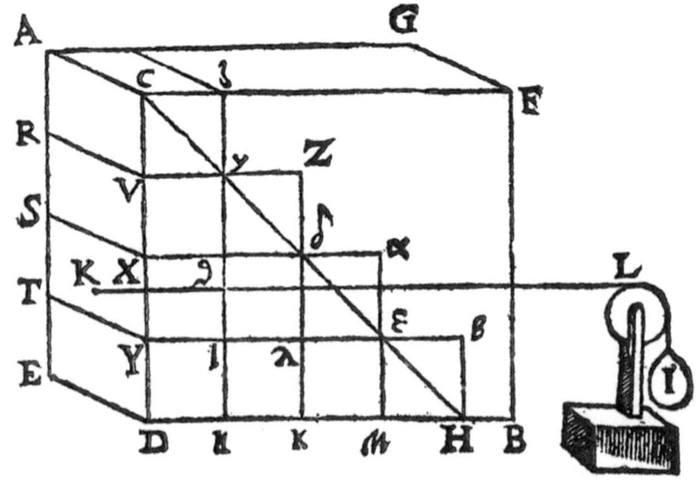

Fig. 3.2 Stevin's diagram for calculating the force on a vetical plane (Dijksterhuis 1955, p. 423)

horizontal planes through RV, SX and TY, so that the vertical plane ACDE is divided into four equal rectangles. To calculate a lower limit to the total force on ACDE Stevin assumes that the degree to which the water presses horizontally is constant across each of the four rectangles and is equal to the degree of pressing at the upper level of each rectangle. By adding the forces on each of the four rectangles calculated in this way Stevin gets a total that is less than the force actually acting on ACDE. Stevin now repeats the process by assuming, once again, that the degree of pressing is constant across each rectangle, but this time it is assumed to be equal to the degree of pressing at the lowest level of each of them. The sum of the four forces gives a result that is greater than the force on ACDE. Stevin now imagines the number of divisions is increased from 4, to 8, 16 and so on and demonstrates that the upper and lower limits of the force on ACDE converge on each other, yielding a result for the force on ACDE that is in conformity with Proposition XI.

We should not be misled either by Stevin's mathematical ingenuity or by the fact that the result he arrives at is correct from a modern point of view. There are two indispensable assumptions implied in Stevin's argument that he does not identify as assumptions and which, moreover, are neither included in nor implied by his postulates. Stevin assumes that the rectangular prism of water does exert a horizontal force on a vertical plane and he further assumes that the force acting horizontally on a vertical surface at some level is equal to the force acting vertically on a horizontal surface of the same area at that level.[12] This latter force is specified by Proposition X, which Stevin was able to invoke. Stevin's postulates say nothing about horizontally acting forces. Postulate III, which I identified above as the only serious contender for the point at which significant content enters into Stevin's hydrostatics, refers only to weights acting vertically downwards and the vertically acting resistance they encounter. Stevin's expression for the force on a vertical plane did not follow from his postulates any more than his expression for the force on a horizontal plane did.

[12] The significance of this point has been missed by commentators. This is illustrated by the way Stevin's theory is treated by Spiers and Spiers in Pascal (1937, p. 150). They included a translation of part of the *Elements of Hydrostatics* as an Appendix to *The Physical Treatises of Pascal* that breaks off immediately before the treatment of the force on a vertical plane that I am arguing to be of great significance. They justify this on the ground that the rest of the work 'consists of interesting geometrical demonstrations whereby Stevin determines the centers of pressure on surfaces variously disposed obliquely to the horizon; its content, not directly relevant to the present discussion'. Since what is referred to as the 'present discussion' is Pascal's introduction of pressure into hydrostatics, Stevin's treatment of the force on non-horizontal surfaces is of crucial importance, as I am endeavoring to show.

3.7 The Significance of the Invalidity of Stevin's Arguments

In this section I spell out the significance of the lapses in deductive adequacy that I have identified in Stevin's hydrostatics, insisting that it goes far beyond mere logical quibbles.

My first point is that when Stevin introduced into his arguments assumptions not included in or implied by his postulates he was violating the standards that he himself invoked. The demand that his theorems be derived from postulates that were sufficiently unproblematic for them to be granted at the outset is Stevin's, not mine, and his introduction of assumptions not licensed by his postulates violated that demand.

My second, more significant, point is that the introduction of unwarranted assumptions by Stevin occurs in precisely those circumstances in which the hydrostatic phenomena in question appear problematic from a point of view restricted to considerations of weight. How can it be that it can take a weight of many pounds to support an ounce of water? How can a body of water press horizontally given that the horizontal component of weight is zero? Given that it does press horizontally, how can it be that the force on a lock gate depends only on the height of the water and not on its extent? Surely, as Stevin's own words implied, these phenomena are 'matters of a more veiled character' that 'might give the critic cause to criticize what does not deserve criticism' and which Stevin must seek to clarify by deriving the relevant phenomena from 'common knowledge that need not be proved'.[13] There is no doubt that the veracity of the assumptions that Stevin did insert into his hydrostatics, whether consciously or unconsciously I do not know, were familiar to him in the context of his hydrostatic practice. Stevin, a hydrostatic engineer in the Netherlands, could hardly have been unaware that water presses horizontally against a dyke or a lock gate. The issue at stake concerns how such knowledge was comprehended and theorized in Stevin's hydrostatics. If there is a puzzle posed by the fact that water presses horizontally against a lock gate and to a degree that is independent of the extent of the water, it does not help, given Stevin's aspirations, to add such knowledge to his postulates as assumptions to be granted.

Stevin did raise the issue of the force on a lock gate in the *Practice of Hydrostatics*.

> As regards the 11th proposition, from it is evident, among other things, what is the weight of the water pressing against either side of the gate of a lock and the like. Also, that the water on one side, even if it were only the width of a straw, exerts the same force against it as the waters having the breadth of the ocean on the other side, provided they are at the same level (Dijksterhuis 1955, p. 497).

It is true that the force on a lock gate follows from Proposition XI, but it does so only because the relevant knowledge has been assumed in its proof. As we have noted, Steven *assumes* that the force on a vertical surface of a given area at some level is equal to the force on a horizontal surface of the same area at the same level.

[13] Here I exploit Stevin's own articulation of what I have referred to as the Euclidean ideal, illustrated in the passages from *The Art of Weighing* cited on p. xx.

It is because that latter force is independent of the lateral extent of the water that the former is too. When Stevin continued the passage quoted immediately above with the words 'Of these matters we do not draw up any special propositions, in view of their aforesaid clearness' he gave the impression that his hydrostatics had clearly explained phenomena associated with lock gates when it fact it had done nothing of the kind.

The fact that Stevin inserted assumptions into his theory that could be argued to be in need of explanation is connected with my third point. Stevin's theory did not provide physical explanations of a number of hydrostatic phenomena. He gave no precise, mechanistic account of how forces are transmitted from one part of a liquid to another thereby bringing about hydrostatic effects. A number of Stevin's successors who built on his hydrostatics found this aspect of it problematic. Isaac Beeckman and Rene Descartes, in 1619, attempted to address what they saw as this deficiency in Stevin's theory by proposing corpuscular mechanisms, but without much success (Descartes 1964–1976, Vol. 10, pp. 67–74). Decades later Robert Boyle (1999, Vol. 5, pp. 207 and 236) complained that Stevin had merely asserted that certain propositions in hydrostatics are true without showing why they are true. By that time, Boyle was able to draw and improve on Blaise Pascal's treatment of the problem, as we shall see in subsequent chapters.

3.8 The Practice of Hydrostatics and Empirical Support

In the *Practice of Hydrostatics* Stevin describes in an informative way a number of practical situations to which his theory of hydrostatics can be applied. It is natural for a modern reader to interpret these ingenious constructions as experiments providing evidence for the theory, and, indeed, they were appropriated and used in that way by Pascal, as we will investigate in Chap. 6. However, Stevin made it quite clear that this was not how he interpreted the situation. The practical situations that he describes are intended by him to be illustrations and applications of, not evidence for, his hydrostatics. From Stevin's point of view, seeking support for his hydrostatics in experiments would have been as inappropriate as doing so for geometry.

The bulk of the *Practice of Hydrostatics* is concerned with some of the counter-intuitive consequences of Proposition X, such as the need for a heavy weight to support a much lighter body of water in the situation depicted in my Fig. 1.1.[14] When introducing his description of the relevant 'experiments' Stevin made his views on their purpose quite explicit.

> We have proved mathematically in the 10th proposition of the elements of hydrostatics, in the fifth corollary [that the force on a horizontal surface depends only on its area and depth and is otherwise independent of the volume of water]. But because many people may consider this unnatural, we will, in addition to the foregoing mathematical proof, describe five

[14] Stevin's version of this diagram appears in Dijksterhuis (1955, p. 492).

practical examples thereof, which anyone may test and see with his own eyes (Dijksterhuis 1955, p. 487).

One 'application' of Stevin's hydrostatics is worth mentioning because it is so straightforward and relatively simple to execute. If one end of a hollow tube is blocked by a smooth metal disc and the arrangement is plunged disc-downwards into water with the upper end of the tube open to the air, then the disc remains suspended in spite of being denser than the water on which it rests.[15]

The stance taken by Stevin on the relationship between his *Elements of Hydrostatics* and practical applications is in line with the metaphor he employed to introduce *The Practice of Weighing* as a sequel to his statics.

> Just as it would be useless to lay large and strong foundations which can support a heavy edifice without ultimately wishing to erect any building thereon, thus in the elements of the arts, theory is lost labor when the end does not tend to practice. Just as, in the natural order of things, the foundations precede the building, thus theory precedes practice. This being so, it is appropriate that, having described the Elements of the Art of Weighing, hereinbefore, I should follow this up with the Practice of Weighing (Dijksterhuis 1955, p. 293).

The foundations of hydrostatics are provided by its 'Elements' that are mathematically proven. Having been established, they can be put into practice in novel as well as familiar situations. To the extent that my appraisal of Stevin's hydrostatics is legitimate then, by expressing himself in this way, he is misrepresenting the relationship between his theory and practice.[16] In my view, Stevin was able to arrive at key theorems in his hydrostatics only by feeding into it knowledge, like the horizontal force on a vertical surface, with which he was thoroughly acquainted as a practicing engineer.

My point can be substantiated more precisely by reference to his dealings with lock gates. Not only was Stevin familiar with lock gates but also he was involved in attempts to improve their design. He devised a swivel-gate lock that was designed to make possible the opening of the gate when the water levels on each side were unequal, so that the resulting surge of water could be utilized for scouring harbors.[17] Stevin's lock gates met to form a V pointing upstream, taking advantage of da Vinci's innovation, mentioned in Sect. 2.1 that was designed to utilize the pressure of the water on the gates to maintain the seal. It is abundantly clear that the technology associated with lock gates presupposed that they were pressed horizontally to a degree that depended on the disparity in heights of the water on either side. As an engineer, Stevin skillfully exploited his knowledge of the way in which water presses against lock gates. As a theoretician he appropriately drew attention to the fact that water presses horizontally and upwards to the same degree that it presses downwards. But there were limits to the extent that Stevin can be said to have theorized that fact. His attempts to impress the burgomasters and rulers of the City of

[15] Stevin's discussion is in Dijksterhuis (1955, pp. 495–497).

[16] The quotation is taken from a dedication of Stevin's *Practice of Weighing* to the burgomasters and rulers of the city of Nuremburg and so should be interpreted while bearing in mind his intent to present his book as serving their interests.

[17] Stevin's work on sluice gates, published in 1617, is discussed in Dijksterhuis (1970, pp. 98–100).

Nuremberg by promising to improve the practice of mechanics and hydrostatics by identifying their theoretical foundations were themselves lacking in foundation.

It would be inappropriate to conclude this discussion of Stevin's hydrostatics on a negative note. Whatever its status from the point of view of logical rigor, the content of Stevin's hydrostatics, expressed by Stevin's law, is correct and contained much that was novel in 1586. Many of the circumstances presented by Stevin as applications in the *Practice of Hydrostatics* drew attention to facts about the behavior of water some of which were both novel and counterintuitive. Over the ensuing decades such facts became accepted as such and, as we shall see, Pascal could exploit them in the 1650s when he came to theorize hydrostatics in a way that differed in significant respects from Stevin's effort.

3.9 Euclidean Explanations and Mechanistic Explanations

The derivation of the theorem of Pythagoras from the postulates of Euclidean geometry serves to establish *that* the theorem is true once the postulates are granted. There is also a sense in which the derivation serves to *explain* the theorem by revealing *why* it is true. A demonstration of how the theorem follows of necessity from accepted axioms serves as what I will term a 'Euclidean explanation' of it. Stevin sought Euclidean explanations of hydrostatic phenomena, and in doing so he followed a norm that was implicit in the long history of the science of weight.

Euclidean explanations were not the only kinds of explanation available to be exploited in Stevin's day. A less abstract mode of explanation is at work in the more common appreciation of the workings of simple machines. Given that a balance arm is rigid and its motion constrained by the pivot, then depression of the weight on one side must result in a lifting of the weight on the other side. The lifting is explained by identifying how the cause, the depression of the former weight, is transmitted to the lifted weight via the mechanism that links them. In the case of the balance the intelligibility of the explanation is a matter of common experience. Explanations of the effects brought about by more complicated machines involving pulleys, gear wheels and inclined planes require more than common experience for their comprehension, especially if those explanations are quantitative as well as qualitative. Nevertheless, explanations of the kind in question involve identification of the way in which causes bring about their effects by way of the mechanisms that link them. I will refer to the kinds of explanation involved here as 'mechanistic explanations'.[18] Mechanistic explanations are intelligible at a common sense level whether the cause is a weight bearing down or a human pulling on the end of a lever. More detailed and precise explanations that moved beyond common sense were possible in the science of weight that focused on and gave a quantitative treatment of weight as a cause. By

[18] I use 'mechanistic' in the context of simple machines and reserve the term 'mechanical' to refer to explanations that appeal to the ultimate particles involved in the mechanical philosophy, thereby hoping to avoid the confusions that can arise from ignoring this distinction.

the time that Stevin turned his attention to hydrostatics, clocks were sufficiently common to serve as the archetype of mechanistic explanations of the kind in question. The cause of the motion of the hands is straightforwardly attributable to the way in which the falling of a weight or the contraction of a spring, is transmitted to them via interconnected gear wheels.

Euclidean explanations presuppose the truth of the postulates or axioms that yield them. Mechanistic explanations have their presuppositions too. Weight is taken for granted in the explanation of machines, as is the rigidity of balance arms, the near constant length of pulley strings, the elasticity of watch springs and the like. Mechanistic explanations are intelligible to the extent that such understandings are indeed taken for granted in everyday dealings with the world. It is true that the science of weight was developed in a technical way that transcended common, everyday, knowledge, when, for instance, the fact that adjacent verticals are not parallel was taken into account or when the notion of center of gravity was formulated and sophisticated mathematical techniques were introduced for locating centers of gravity of regular geometrical shapes. But even the most advanced and technical aspects of such developments did not run counter to the fact that physical properties presupposed in mechanistic explanations, such as weight, elasticity and rigidity, were appealed to in such explanations though not themselves explained. As we shall see in the course of this book, the legitimacy or otherwise of explanations that appeal to unexplained causes was a contentious issue with many mechanical philosophers. Such contention aside, it is a matter of historical fact that the science of weight available to Stevin did give mechanistic explanations of a range of mechanical phenomena and such explanations were straightforwardly intelligible and facilitated purposeful intervention in the world.

If we turn a blind eye to the invalidity of key deductions in Stevin's *Elements of Hydrostatics* then we can acknowledge that he gave a Euclidean explanation of the phenomena by deriving postulates that described them. However, this fell short of giving a mechanistic explanation of those phenomena. If what is wanted is an account of the pushes and pulls that link cause to effect and which serve to explain the phenomena thereby produced, then Stevin's hydrostatics did not fully provide them. Stevin did not offer an account of how forces are transmitted through liquids analogous to the way in which they are transmitted via balance arms, pulley strings or gear wheels in a simple machine. He could not give an account of how it is that water can push vertically upwards on a floating body and horizontally against a lock gate. A significant number of those who sought in the seventeenth century to build on Stevin's pioneering efforts saw this as a deficiency that they hoped to remove, with varying degrees of success.

The shortcomings of Stevin's hydrostatics as far as its explanatory capacities are concerned can be illustrated quite strikingly with reference to one of the five 'practical examples' that he had to offer in the *Practice of Hydrostatics*. A cursory glance at the figure accompanying Stevin's discussion, reproduced as Fig. 3.3, might lead one to suspect that he is concerned with what was generally regarded as a paradox in need of explanation. How can the small amount of water in the tube on the left support the much larger amount of water in the tube on the right? This is not the

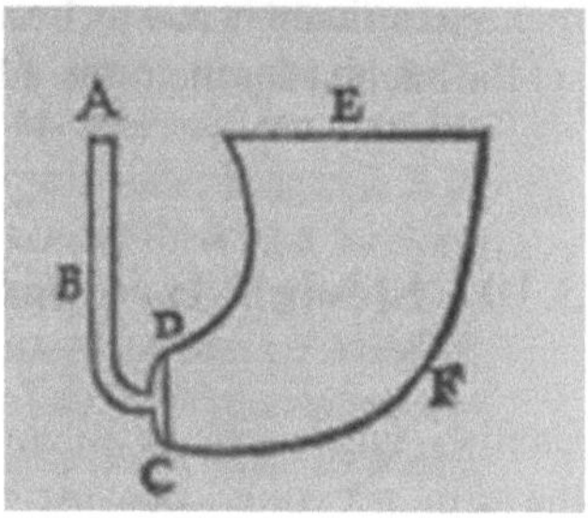

Fig. 3.3 A small volume of water balancing a larger one (Dijksterhuis 1955, p. 490)

question addressed by Stevin in his text. Rather, he is concerned with establishing that the force exerted by the water on the 'bottom' (that is, 'solid surface') CD is the same on each side. I quote Stevin's argument in full.

> Now that the large water of the vessel CDEF does not exert any greater pressure against the bottom CD than does the small water of the small tube, appears in practice as follows: Let the bottom DC be taken away; then the large water will thrust against the smaller in that place. Now if the water CDEF had previously exerted a greater pressure against the bottom CD than the water ABCD, it will now also exert a greater pressure against the former water than that against this, as a result of which that which is weaker will have to yield to that which is stronger, i.e. the water ABCD will ascend and that of CDEF will descend. But this being so, their upper surfaces will not be on a level, which is manifestly contrary to experience. The smaller water ABCD therefore exerts the same pressure against the bottom CD as the larger water CDEF.

The theoretical assumption put to use in the argument so formulated is 'that which is weaker will have to yield to that which is stronger'. That statement is virtually identical to what was in effect Archimedes' attempt to capture the fluidity of liquids. Stevin does not invoke any of the details of his own hydrostatics that go beyond Archimedes. The other assumption is an empirical one. The water must rest at the same level in each of the tubes because to deny that 'is manifestly contrary to experience'. There is certainly every reason to grant Stevin his empirical fact. After all, it had been known for over 2000 years that water in a conduit that brings water to a city from a high-altitude lake cannot be made to rise above the level of the water in the lake. What is at issue here is not that the water in the higher region of the conduit rests at the level of the water in the lake but the puzzle posed by this undoubted fact. How can the relatively small amount of water in the conduit support the weight of the vast quantity of water in the lake? Stevin had little to offer by way of an *explanation* of that circumstance.

In retrospect, it can be said that what was needed was an account of the way in which mechanical systems are constrained by the presence of a body of liquid, an issue that takes on a special significance once it is recognized that the constraints imposed by the parts of the simple machines all depend on the solidity of the parts. The mechanical action of liquids could be understood only when the nature of fluidity was captured via a technical concept of pressure up to the task. We will see that Pascal, Boyle and Newton improved hydrostatics at least in part by contributing to the development of the concept of pressure, whereas Galileo and Descartes did not make significant headway because their efforts lacked such a feature. These remarks set the agenda for the remainder of this book.

3.10 Stevin's Terminology and the Concept of Pressure

English translations of Stevin's *Elements of Hydrostatics* tend to make free use of the term 'pressure'. Because of this a reader is likely to read into Stevin's text more of what is implied by the modern notion of pressure than is justified. Stevin did occasionally use the noun '*drucking*', the sixteenth-century Dutch equivalent of 'pressure' as well as derivatives of the verb '*drucken*', to press. But he also made extensive use of other words such as '*beswoort*' and '*gheprangt*', which both roughly translate as a kind of pushing or pressing and the nouns '*anstoot*' *and* '*ghe-prangselen*' to describe the results of such pushing or pressing. None of these terms are introduced as technical ones by Stevin. He drew on a range of non-technical terms available from the language of his day to give expression to his claims about the way in which water presses on various solid surfaces. When Dijksterhuis translates this range of terms using the single term 'pressure' he obscures this aspect of Stevin's text and gives it a modern ring that is not justified.

I will illustrate these general remarks with some examples.

At the very beginning of the *Elements of Hydrostatics* Stevin summarized the argument that he was about to present. The single paragraph in which he did so contains the following sentence.

> The 10th, 11th, 12th, 13th, 14th, 15th propositions are to deal with the force of the pressure (*drucking*) of the water against surfaces.

When we turn to Proposition X and the ensuing discussion we find that the term *drucking* is not used. The force on a horizontal surface is attributed to the fact that there rests a weight (*rust een ghewicht*) on it. This terminology is repeated when the discussion moves to the water pressing on the planes EF in the variously shaped tubes of water depicted in Fig. 3.2, even when the water is pressing up on EF as in the third of the three figures. When Stevin turns to the horizontal force on a vertical plane in Proposition XI he again talks in terms of weight resting against the surface (*T'ghewicht daer teghen rustende*) rather than in terms of pressure. Again, the term *drucking* is not employed (Dijksterhuis 1955, p. 421). When Stevin refers to what Dijksterhuis translates as the 'center of gravity of the force of the total pressure on the surface' he uses *gheprangs* for what Dijksterhuis has translated as 'pressure'. The common usage of that term typically referred to the results of a number of people pushing together, as in the force exerted by a surging crowd.

There is a passage at the end of Stevin's *Practice of Hydrostatics* in which he offers an explanation of why a man under water is not troubled by the force of the water above him. This, in Dijksterhuis' translation, is due to the fact that 'the water exerts the same pressure on all sides'. This translation is quoted by Alan Shapiro (1974, p. 262) in support of his case that Stevin had a grasp of fluid pressure way ahead of his contemporaries. However, such an interpretation becomes less plausible once it is realized that Stevin did not use the single term '*drucking*' in this context but a number of alternatives as well. He used '*anstoot*' not '*drucking*' for the force that presses a man under water from all sides ('*ouermits t'water an allen sijden euestijf anstoot*'). What is more, the details of Stevin's explanation of why the man under water is not crushed to death appeals to the fact that water does not thrust when in its natural place. When he deals with the fact that the water presses slightly more on the top of the man than on the bottom, Stevin talks of that pressing using the term '*gheprangt*'. The fact that Stevin uses the common words 'drucking', 'anstoot' and 'gheprangt' somewhat indifferently to refer to the pressing of water is a clear sign that he is treating it as a common concept presumed in the language of his day rather than a technical one.

There is no technical notion of pressure in Stevin's hydrostatics. When he did talk of water pressing against a solid surface he did so by using a number of terms available to him from common usage. He did occasionally use '*drucking*' but in a way that gives no grounds for seeing in it as an anticipation of the technical meaning that later generations were to give to that term.

As we have seen, Stevin did use *bodem* (bottom) as a technical term. When he took the step that recognized that water can press horizontally as well as vertically, his description of it in terms of water weighing against bottoms appears uncontentious when 'bottom' is given its common sense meaning, but is much less so once the technical meaning is embraced. This gradual move from common sense knowledge of weight to other kinds of force that could be explicated employing analogies with weight was typical of the time. The pressing of liquid against a vertical surface is a force analogous to weight and can even be measured against weight. In this connection it is relevant to reflect on the function of the weight *I* shown in Fig. 3.2 that balances the force on plane ACDE and is sufficient to hold that plane in place should it be movable. The weight acts as a measure of the force on the plane and helps to make talk of the water weighing on the plane appropriate. Of course, there is no question of Stevin actually performing such an experiment, nor is it in fact performable so long as there is no way of arranging for the horizontal part of the string attached to the weight to pass through the solid surface GFB is given.

3.11 Where to from Here?

The content of Stevin's hydrostatics, conveniently summarized in what has become known as Stevin's law, was both novel and important at the time Stevin gave expression to it. I have argued above that the main reason why Stevin was able to latch on to his important insights stemmed from his familiarity with the behavior of water gained through his practice as an hydraulic engineer. He aspired to do more than that. Stevin aimed to present his hydrostatics as a mathematical theory following the style and standards implicit in works of Euclid and Archimedes and instantiated in the science of simple machines that he himself advanced in his *Art of Weighing*.

Stevin's hydrostatics did not fully live up to his aspirations. He did not formulate postulates from which Stevin's law follows, so his attempts to provide Euclidean explanations of hydrostatics failed. Nor he did not give mechanistic explanations of hydrostatics effects in the way that the science of weight provided mechanistic explanations of the simple machines.

The postulates of Stevin's hydrostatics were too closely tied to considerations of weight and failed to give theoretical expression to the way in which liquids differ from solids. Connected with this is the fact that he was not able to give an account of how hydrostatic causes are linked to their effects by way of the transmission of pushes through or by liquids. He did not explain how hydrostatic forces are able to bend around corners. In retrospect it can be appreciated that what was required was the recognition of the action of pressure as a cause in addition to weight.

There was no rival account of hydrostatics in Stevin's lifetime that provided what I have argued he failed to provide. The path to an adequate theoretical account of hydrostatics that was able to grasp theoretically the difference between solids and liquids and supply a causal, mechanistic account of hydrostatic effects was by no means obvious. This much is apparent from the limitations of the attempts by Stevin's successors to advance hydrostatics. In the next two chapters I illustrate this point by looking at the attempts by two major figures, Galileo and Descartes, to do that. Then, in Chap. 6 I turn to the hydrostatics of Pascal that represents the first significant advance towards the technical concept of pressure, 68 years after Stevin's pioneering efforts.

References

Boyle, R. 1999. *The works of Robert Boyle*. 14 volumes, ed. M. Hunter and E. Davis. London: Pickering and Chatto.

Dear, P. 1995. *Discipline and experience: The mathematical way in the scientific revolution*. Chicago: University of Chicago Press.

Descartes R. 1964–1976. *Ouevres de Descartes*. 12 volumes, ed. C. Adams and P. Tannery. Paris: Vrin.

Dijksterhuis, E.J. 1955. *The principal works of Simon Stevin, Volume 1, Mechanics*. Amsterdam: Swets and Zeitlinger.

———. 1970. *Simon Stevin in the Netherlands around 1600*. The Hague: Matinus Nijhoff.

Duhem, P. 1905. Le Principe de Pascal. *Revue Générale des Sciences Pures et Applique*. 16: 599–610.

Gaukroger, S., and J. Schuster. 2002. The hydrostatic paradox and the origins of Cartesian dynamics. *Studies in History and Philosophy of Science*. 33: 535–572.

Pascal, B. 1937. *The physical treatises of Pascal: The equilibrium of liquids and the weight of the mass of the air*. Trans. A.G.H. Spiers and I.H.B. Spiers. New York: Columbia University Press.

Schemmel, M. 2008. *The English Galileo: Thomas Harriot's work on motion as an example of pre-classical mechanics*. Dordrecht: Springer.

Shapiro, A. 1974. Light, pressure and rectilinear propagation: Descartes' celestial optics and Newton's hydrostatics. *Studies in History and Philosophy of Science*. 5: 239–296.

Chapter 4
Galileo, Floating Bodies and the Balance

Abstract A natural way of interpreting Archimedes' account of floatation is by analogy with a balance. When a body is wholly or partially immersed in water it displaces water, which is raised as a result. Floating occurs when the tendency of the raised water to move downwards is balanced by the weight of the body. Galileo came to realize that if this comparison with the balance is to be retained the analogy must be with a balance with unequal rather than equal arms. If a balance with unequal arms is subject to a small displacement about its equilibrium position the resulting velocities of the bodies at the extremities of each arm are inversely proportional to their weight. Likewise, a slight vertical depression of a block of wood floating in a vessel will result in a displacement of water such that the velocity of the displaced wood or water is inversely proportional to its weight. This analysis yields Archimedes principle. However, it is confined to considerations of weight and of vertical displacements, as the approach of Archimedes had been, and was not well-equipped to deal with the isotropic hydrostatic forces noted by Stevin.

4.1 The Context of Galileo's Account of Floatation

Galileo discussed floatation in an unpublished early work, *De Motu*, probably composed when he was at the University of Pisa from 1589 to 1592. The discussion is set in the context of an Aristotelian concept of natural motion of the elements that Galileo largely accepts but also amends in important respects. This early account of floating drew on analogies with the equal armed balance and clearly owed a debt to Archimedes as well as to Aristotle.[1]

Galileo came to find inadequacies in the Archimedean theory of floatation and eventually replaced it with one that was published in 1612 in his *Bodies that Stay*

[1] Galileo's early account of floating appears in manuscript notes of Galileo's gathered under the title *De Motu,* probably completed during his time at the University of Pisa. An English translation of that work from the Latin, as well as of *Le Meccaniche*, written early in Galileo's Paduan period, appears in Drabkin and Drake (1960). Galileo's early engagement with floating appears on pp. 16–23. Galileo's debt to Archimedes with particular reference to hydrostatics is documented in Shea (1972, Chapter 1) and in Palmieri (2005).

A.F. Chalmers, *One Hundred Years of Pressure*, Archimedes 51,
DOI 10.1007/978-3-319-56529-3_4

atop Water or Move in It.[2] The mature theory drew an analogy with a balance constructed with unequal rather than equal arms. In this chapter I characterize Galileo's mature theory, noting its relation to and improvement on the Archimedean theory that he initially adopted, and then comment on some of its general features.

At the time he wrote *Bodies that Stay atop Water or Move in It*, Galileo had for 2 years been in Florence in his position as mathematician and philosopher to Cosimo Medici, the Grand Duke of Tuscany. He acquired his position with the help of the fame he had established through his telescopic observations of the heavens, including the discovery of the moons of Jupiter that he had diplomatically named the Medicean planets. As Galileo explained at the beginning of his book, the writing of it was somewhat of a diversion from the astronomical project that had become his major preoccupation and which was eventually to lead to the publication of the *Dialogue Concerning the Two Chief World Systems* in 1632. What had seemed to be a minor discussion about the floating of ice and ebony chips in or on water had turned into a major dispute between Galileo and his Aristotelian rivals. This was because Galileo's position ran counter, for example, to some basic Aristotelian assumptions about the four elements and the motions natural to them. Also in dispute were questions about whether shape is an essential factor involved in floatation and whether liquids offer resistance to motion. One result of the detailed, and occasionally heated, dispute was that Galileo came to recognize that he needed to revise what was basically his Archimedean account of floatation.

Galileo's *Treatise on Floating Bodies*, as I will refer to *Bodies that Stay atop Water or Move in It*, was written in Italian and designed to engage a wide audience. So successful was it in that respect that the first edition sold out within a month. Galileo hastily prepared a second edition, also published in 1612. Galileo had refined his views in the intervening period and incorporated additions and improvements into the new edition. The evolution of Galileo's views on hydrostatics from 1611 to 1612 has been reconstructed from manuscript material by Paulo Palmieri (2005). I do not attempt to reproduce the details here. My main concern is with Galileo's considered views as they appear in the second of the two 1612 editions of the *Treatise on Floating Bodies.*

The transformation in Galileo's views on hydrostatics, analyzed in the remainder of this chapter, can be summarized succinctly as follows: In his *Treatise on Floating Bodies* Galileo replaced his account of floating as it appeared in *De Motu,* which was based on an analogy with a balance with equal arms, with a new account employing an analogy with a balance with unequal arms.

[2] I have used the translation of Galileo's work on floating bodies in Drake (1981, pp. 18–53). This translation includes the additions that Galileo made in the second edition.

4.2 Floating Compared with a Balance with Equal Arms

Galileo's discussion of bodies in water in *De Motu* was couched in terms of Aristotelian notions of heaviness and lightness and the naturalness of the motions to which they give rise, a standpoint that he was to move away from in subsequent work. The account is closely reminiscent of Archimedes' reasoning to which Galileo explicitly added an analogy with the balance.

Chapter 3 of *De Motu* has the title 'That natural motion is caused by heaviness and lightness'. The ensuing discussion makes clear that the motions in question are that of solid bodies through a fluid medium and that 'heaviness' and 'lightness' refer to the weights of equal volumes of the bodies or media in question, that is, to their densities. Galileo takes for granted a distinction between solids and liquids insofar as he recognizes that liquids freely give way to the passage of solids through them. The natural motions in question are those that involve solids of various densities freely rising or falling through media of various densities.

Chapter 4 is headed 'First Demonstration: in which it is proved that bodies of the same heaviness as the medium move neither upward nor downward'. The proof goes as follows. If part of a body equal in density to water is immersed in water then it will displace a volume of water equal to the volume immersed. As a consequence 'the level of the water is necessarily raised'. The displaced water 'strives by its weight to return downwards to its original position' but cannot achieve this unless the immersed solid is lifted. If equilibrium is to obtain, 'the weight of water—by which the water strives to raise the solid upward, must necessarily be equal to the weight with which the solid resists and presses downwards' (Drabkin and Drake 1960, p. 18). Given that the solid and the water have the same density it follows that equilibrium will be established when the top surface of the immersed solid coincides with the surface of the water.

In Chapter 5 Galileo proves that 'bodies lighter than water cannot be completely submerged'. Were a solid less dense than water to be completely submerged it would displace, and therefore raise, a volume of water equal to its own volume. The displaced water will strive downward by its weight, and so strive to lift the solid. Since the water is denser than the solid it will strive to lift the solid to a degree that exceeds the weight of the solid. So the solid will not remain totally submerged but will be lifted. In a similar fashion Galileo shows that solids more dense than water will move downward when totally immersed in it.

Galileo's proofs are very close to those of Archimedes with the only difference being that Galileo considers objects immersed in water enclosed in vessels whereas Archimedes has considered objects immersed on a spherical ocean. As we will see, and as Galileo came to realize, this move in fact had the effect of transforming Archimedes' valid proofs into invalid ones.

Galileo did not rest content with the above treatment. As he explained:

> But the points set forth in these last two chapters cannot very well be further elucidated mathematically; they require rather a physical explanation. For this reason I propose, in the next chapter, to reduce the matter to a consideration of the balance, and to explain the

> analogy that holds between bodies that move naturally and the weights of the balance. My aim is a richer comprehension of the matters under discussion, and a more precise understanding on the part of the reader (Drabkin and Drake 1960, p. 20).

A weight on the end of a balance will have a natural tendency to move downwards. That tendency can be countered, and the weight forced upwards, if a greater weight is suspended from the other, equal, arm of the balance. From this consideration Galileo extracted the general principle that 'the heavier cannot be raised by the less heavy' (Drabkin and Drake 1960, p. 21). The reasoning could then be transferred to bodies immersed in water.

> If, for example, we let a piece of wood down into water, then, if the wood is to be submerged, water must necessarily leave the place into which the wood enters, and this water must be raised, that is, must be moved in a direction away from the center of the universe. If, then, the water which has to be raised is heavier that the wood, it surely will not be able to be raised by the wood. But if the whole piece of wood is submerged, then from the place into which the wood enters a volume of water must be removed equal to the volume of the whole piece of wood. But a volume of water equal to the volume of the wood is heavier than the wood (for it is assumed that the wood is lighter than water). It will therefore not be possible for the wood to be completely submerged. And this is in agreement with what was said in the case of the balance, namely, that a smaller weight cannot raise a larger.

In comparing the support of the weight of floating bodies by the weight of the liquid displaced to the way in which weights mutually support each other on an equal-armed balance Galileo was following a line of reasoning that was strongly suggested by Archimedes' treatment of floatation. This is a point that Palmieri (2005, p. 199) has stressed and which I noted in the discussion of Archimedes theory in Sect. 2.3. However seductive this way of thinking may be, it is defective in a way that Galileo came to realize. It is not in general true that if a volume of solid is immersed in a liquid an equal volume of liquid is raised.

4.3 The Inadequacy of the Analogy with an Equal-Armed Balance

When Galileo turned his attention to flotation in 1611 the sources on which he drew were Archimedes' works on the balance and floating bodies and the Aristotelian analysis of the lever in *Mechanical Problems*. There is no evidence that he was familiar with Stevin's *Elements of Hydrostatics* that had been translated into Latin by 1608.[3]

Galileo came to realize that the comparison of floating with a balance with equal arms, according to which the weight of the floating object is balanced by the weight

[3] This point has the support of Stillman Drake (1981, p. 214), who observes that 'technical mathematical books were little in demand in Italian universities then, and after Galileo moved to the Florentine court in 1610 he probably had even less access to such books published abroad than when he taught in Padua'.

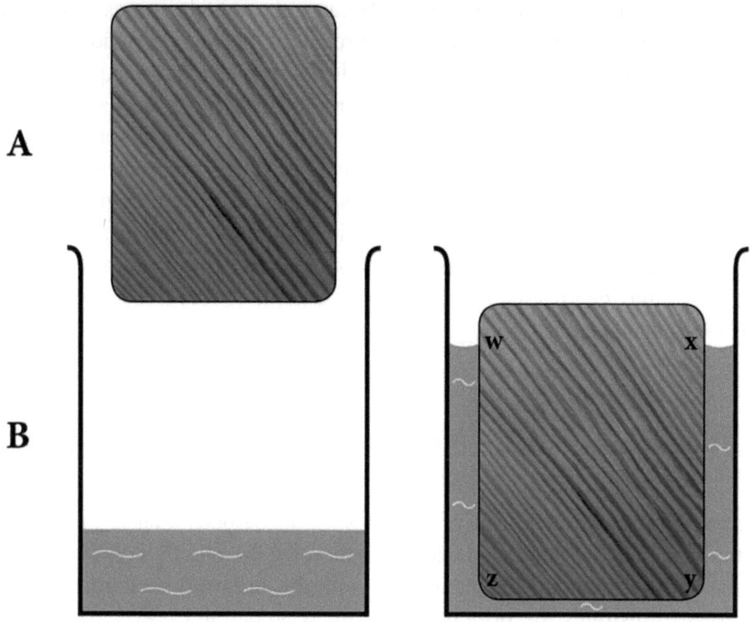

Fig. 4.1 A block of wood can float in a volume of water weighing less than itself

of the water it displaces, is untenable. As Galileo explained when introducing his discussion of floatation in the *Treatise on Floating Bodies*:

> We might say that those [solids] float that are less in specific weight than water, and those sink that are greater in specific weight, inasmuch as it appears that a solid in submerging will raise as great a volume of water as that part of its own body that is submerged, whence it would be impossible for a body of less specific weight than water to submerge itself entirely, it being unable to lift a greater weight than its own, as would a volume of water equal to it in bulk.—Yet this business proceeds otherwise, and though the conclusions are true, the causes thus assigned are defective, nor is it true that the submerging solid thrusts aside and raises a volume of water equal to its own submerged bulk. (Drake 1981, p. 33)

The analogy with a balance with equal arms is inappropriate because the water raised by a submerged body 'is always *less* than the submerged portion of the solid'. This somewhat counterintuitive fact can be simply illustrated. Imagine a solid, A, less dense than water and with a cross-section only slightly smaller than that of a vessel, B, into which it is to be inserted, as illustrated in my Fig. 4.1. Imagine further that, prior to insertion, a small amount of water, weighing much less than the solid to be inserted, lies at the base of the vessel. When the solid is inserted it displaces water that rises up around the solid until the situation depicted on the right of the figure eventuates. Here wxyz marks the cross section of the volume of solid that is beneath the new water surface. The solid floats when the weight of water equal in volume to the volume of solid with cross section wxyz is equal to the weight of the solid. The weight of the solid is greater than the weight of water that has been displaced. Indeed, it is greater than the weight of water available to be displaced!

The counterintuitive conclusion arises because the solid becomes submerged not only as a result of it being lowered, but as a result of the water around it rising. This circumstance is not apparent if the solid is lowered into a large expanse of water, such as the spherical ocean considered by Archimedes, because in such a situation the rising of the water is negligible and the volume of water displaced is indeed equal to the volume of solid submerged. This is no longer true when floating takes place in finite-sized vessels. In such cases, as Galileo came to recognize, 'the water raised is always *less* than the submerged portion of the solid, and proportionally less as the vessel containing the solid is narrower' so that 'a huge beam weighing, say, 1000 pounds, may be raised and sustained by water that does not weigh 50 pounds' (Drake 1981, p. 34).

4.4 Floatation Compared to a Balance with Unequal Arms

The recognition that water in a vessel rises around a body that is inserted into it faster than the body descends led Galileo to theorize floating by drawing an analogy with a balance with unequal arms, that is, with the steelyard. According to Galileo the steelyard illustrates the principal that 'the moment and power of heaviness is increased by speed of motion, so that absolutely equal weights conjoined with unequal speeds are of unequal power, moment, and force, in the ratio of one speed to the other'. In the case of the balance with unequal arms:

> [the weights] do not press and exert force equally, but that which is the greater distance from the center (about which the balance moves) goes down, raising the other; and the motion of that which rises is slow while the other is swift. Now, the power and force conferred by [greater] speed of motion on the moveable receiving it is such that this can exactly compensate the additional weight in the other, slower, moveable.—This equalization between weights and speeds is found in all mechanical instruments and was considered as a principle by Aristotle in his *Questions of Mechanics* (Drake 1981, p. 31).

Galileo here adopts what I will refer to as the principle of inverse proportionality, acknowledging its source in the Aristotelian tradition and applying it to bodies partially or wholly immersed in water in an original way. If we imagine a body floating in a vessel to be slightly depressed from its equilibrium position then the water around it will rise. The inverse proportionality principle requires that the weight of the solid lowered times its velocity must equal the weight of water raised, times its velocity. Application of this principle enabled Galileo to prove a number of facts about floatation.

One result was, in effect, his version of Archimedes' principle. According to it 'solids of less specific weight than water submerge only to a point such that water in the same volume as the submerged part will weigh as much as does the entire solid'. (Drake 1981, p. 47) This can be compared to Archimedes' Proposition 5:

> Any solid lighter than a fluid will, if placed in the fluid, be so far immersed that the weight of the solid will be equal to the weight of fluid displaced. (Heath 1950, p. 257)

Water equal in volume to the submerged part of an immersed solid is not the same as the volume of water displaced by the solid, and it was through his appreciation of this that Galileo was able to replace Archimedes' problematic expression of the conditions necessary for floating with a more adequate one. He was able to accommodate the change by replacing the analogy with the balance with equal arms, which is undermined by the new recognition, with an analogy with a balance with unequal arms using the inverse proportionality principle.

Another result proved by Galileo was that the level at which a body floats in water is independent of the extent of the water by which it is surrounded. He compared two situations in each of which a body floats in water of a given depth but in vessels that differ greatly in cross-sectional area. When the floating body is depressed slightly from its equilibrium position, the water level in the wider vessel rises less quickly than in the narrower vessel because of the greater cross-section of the former. On the other hand, more water is displaced in the case of the wider vessel since it is closer to an infinite expanse of water, in which circumstance the displaced weight would be equal to that of the floating solid, and the displaced water rises at the same speed as the immersed body descends. Galileo was able to apply the inverse proportionality principle to show that the level at which floating occurs is independent of the size of the vessel, even if the latter were to exceed that of the floating solid only by a hair's breadth (Drake 1981, p. 48).

4.5 Some Reflections on Galileo's Account of Floatation

Galileo's introduction of the principle of inverse proportionality of weights and speeds in *Le Meccaniche* followed his derivation of the equilibrium conditions for a balance with unequal arms. This was itself an extension of considerations of the balance conditions for equal arms that exploited the notion of center of gravity. He introduced the new topic as follows:

> Having shown how the moments of unequal weights are equalized by being suspended at distances having the same ratio, it seems to me that we should not pass by in silence another agreement and probability by which the same truth may reasonably be confirmed to us. (Drabkin and Drake (1960, p. 155).

Galileo proceeded to show that the inverse proportionality of weights and the speeds with which they move when a balance is displaced about its equilibrium position is a consequence of the specification of the equilibrium conditions for a balance with unequal arms. What Galileo here describes as 'an agreement and a probability', construed as such by being a consequence of the balance conditions for a steelyard, is treated by him in the *Treatise on Floating Bodies* of 1612 as a *principle* of general application. This was an innovative move that opened up possibilities of further applications of the inverse proportionality principle and was eventually to contribute to a path that led to the concept of energy and its conservation.

Galileo pinpointed a shortcoming in an understanding of Archimedes' account of floatation by analogy with the balance and showed how the analogy could be preserved by invoking a balance with unequal rather than equal arms. However, there is a need to qualify the extent to which that move allowed him to explain floating. Galileo was able to show how the equilibrium conditions in the two sets of cases followed from the inverse proportionality principle, but that falls short of grasping what it is that gives rise to the respective phenomena. There are grounds for claiming that Galileo did not explain floating.

Galileo made a point of proving that the height at which a body floats in a vessel is independent of the dimensions of the vessel and hence of the amount of water that surrounds the floating body and presses on it, as we saw in the previous section. Galileo's proof via the inverse proportionality principle relies on the fact that while more water is raised by a given depression of a solid when the cross-section of the vessel is increased, this increase in the amount of water raised is offset by the slower speed at which it is raised. Galileo's explanation of any particular instance of floating makes an appeal to the ratio between the cross-section of the floating body and that of the vessel in which it floats. But the fact that a body floats and the depth at which it does float turns out to be independent of the cross-section of the vessel! There is something wrong here. It had long been common knowledge that wood floats in water because it is less dense than water, irrespective of the dimensions of the vessel in which it floats. The vessel might well be as large as an ocean. An explanation of floatation, such as Galileo's, that invokes the cross-section of the vessel in any particular instance of floating, cannot be getting at its root cause. Galileo did not give a mechanistic account of floating insofar as he did not identify the force that counters the weight of a floating body in a way that went beyond what was common knowledge at the time.

In the second version of the *Treatise on Floating Bodies* Galileo took his deliberations beyond floating. He showed how the inverse proportionality principle can be used to explain the fact that liquid in a narrow tube can support the greater amount of water in a wider tube to which it connected (Drake 1981, pp. 49–50). His treatment yielded the correct result, namely, that in equilibrium the two columns of liquid have the same height. However, he did not emulate here the *mechanistic* explanation of the balance of a steelyard that appeals to the interaction of the weights via the balance arm. In particular, he did not explicate how it is that one column of water, pressing down, can result in an upward force on the other. He simply takes that as a common-sense fact.[4]

As its title indicates, Galileo's *Treatise* was concerned with floatation and not with hydrostatics in general, so it would be out of place to criticize Galileo for not providing a theory of the latter. Nevertheless, some reflections on what Galileo's treatment of floating involved and could and could not provide does help set the scene for reconstructing and understanding the developments in hydrostatics that

[4] The fact that water rises up one arm of a u-tube when pressed down in the other is common sense only because the capacity of liquids to flow is taken for granted. The same thing does not happen if the water is replaced by sand.

lay ahead. Like the hydrostatics of Archimedes and Stevin, Galileo's account of floatation involved a focus on weight. Galileo could help himself to theories of statics that afforded a sophisticated treatment of weight by way of concepts such as center of gravity. But an appeal to weight is insufficient for grasping the behavior of liquids as opposed to solids. What Galileo did was what both Archimedes and Stevin before him had done. In an *ad hoc* and opportunist way he introduced into his treatment of liquids various assumptions drawn from common sense. For example, he took for granted that water presses upwards on solids immersed in them and that water pressed down on one side of a u-tube rises up on the other. These and associated phenomena, like the pressing of water horizontally against a lock gate and the leaking of water from the sides and top as well as the base of a pipe, pointed to features of the action of liquids which were to require a grasp of the distinction between solids and liquids that necessarily went beyond considerations of weight and the common sense realization that liquids flow in a way that solids do not. As we will see, it was Blaise Pascal and Robert Boyle who made significant moves in that direction. Before investigating them I add a chapter on another way of dealing with liquids that, as far as hydrostatics is concerned, proved to be more significant through its shortcomings rather than its successes. I refer to the hydrostatics that can be extracted from the writings of Rene Descartes.

References

Drabkin, I.E., and S. Drake. 1960. *On motion and on mechanics*. Madison: University of Wisconsin Press.

Drake, S. 1981. *Cause, experiment and science: A Galilean dialogue incorporating a new English translation of 'Bodies that stay atop water or move in it'*. Chicago: University of Chicago Press.

Heath, T.L. 1950. *The works of Archimedes*. New York: Dover.

Palmieri, P. 2005. The cognitive development of Galileo's theory of buoyancy. *Archive for History of Exact Sciences*. 59: 89–222.

Shea, William. 1972. *Galileo's intellectual revolution*. London: Macmillan.

Chapter 5
Descartes' Engagement with Hydrostatics

Abstract The young Descartes was involved in contributing to the sciences of his day, especially mechanics and optics and including a brief, unpublished, foray into hydrostatics. Such involvements persisted into his maturity, but came to play a role subsidiary to Descartes' ambitious plan to construct an outline of the ultimate, corpuscular, structure of the universe in its entirety. Descartes' contributions to hydrostatics need to be extracted piecemeal from his writings because he did not offer a systematic treatment. In the early, unpublished, work Descartes did respond to some of Stevin's results in a way that can be interpreted as having potential, but he did not pursue those early thoughts. The corpuscular system of the world outlined in Descartes' *Principles of Philosophy* treated pressure in liquids and the transmission of light in analogous ways. Both were seen as resulting from the transmission of pushes from corpuscle to corpuscle. This had the result that the linear transmission of hydrostatic forces in liquids was as fundamental as the linear transmission of light. In retrospect at least it can be seen that Descartes' hydrostatics was ill-equipped to deal with the isotropy of hydrostatic forces that had been identified by Stevin.

5.1 Introduction

Apart from an early, unpublished, manuscript there is no work of Rene Descartes that was devoted specifically to hydrostatics. Consequently, the following account of his contributions to that topic has been assembled from a diversity of sources. As well as the early manuscript they include his engagement with the mathematical sciences of his day, especially optics, mechanics and astronomy, his attempt to construct a mechanical account of the universe in its entirety in *Le Monde* and his *Principles of Philosophy*, and his correspondence. After some scene setting I give an account and appraisal of Descartes' hydrostatics as it appears in his early manuscript and follow this with the, rather different, treatment of hydrostatics that resulted from Descartes' adaptation of the work of Archimedes. My third topic is the relationship between Descartes' hydrostatics and his mechanical system of the Universe. As a preliminary to examining that, it is necessary for me to review Descartes' involvement with light, which played a central role in his mechanical picture of the world and the principles governing it.

A.F. Chalmers, *One Hundred Years of Pressure*, Archimedes 51,
DOI 10.1007/978-3-319-56529-3_5

The general tenor of my assessment of Descartes' hydrostatics is that it was unproductive. The treatment he developed in the early manuscript was original and had potential, but that potential was not pursued by Descartes and the fact that the manuscript was never published ensured that it was not pursued by others either. Insofar as Descartes adopted a common sense distinction between liquids and solids and Archimedes' account of floating bodies he inherited the limitations inherent in them. In particular, Descartes did not develop them in a direction that would help to accommodate the fact, implicit in Simon Stevin's law, which specifies that hydrostatic forces act equally in all directions. That is, Descartes did not accommodate the isotropy of hydrostatics forces. This particular drawback was entrenched rather than alleviated by the details of the way in which Descartes incorporated hydrostatics and optics into his mechanical system, as we shall see. In that context, Descartes likened the transmission of hydrostatic forces through a liquid to the transmission of light. As a consequence, the understanding of hydrostatic forces as acting only linearly downward was as necessary as, and was intimately linked with, the recognition that light travels in straight lines.[1]

5.2 Physico-mathematics and the Mechanical Philosophy

Descartes' early work involved extensions of traditional sciences such as the science of weight, geometrical optics, planetary astronomy and the impetus theory of motion with which he had become familiar in the latter years of his education at the Jesuit collage, *La Flèche*. Later, Descartes attempted to transcend those sciences by reducing them to his mechanical account of the universe involving nothing other than portions of matter/extension in motion. For example, conceptions such as weight or light rays figuring in the traditional science of weight and geometrical optics respectively were explained in the *Principles*. The category of 'mixed mathematics' into which the traditional sciences were situated by Aristotelians made sense insofar as physical explanations (drawing, for example, on the concept of natural motion) could be distinguished from considerations involving an appeal to mathematics that could be abstracted from physical considerations and subject to a deductive treatment, Euclidean-style. Insofar as basic processes, such as the impact of colliding bodies for instance, came to be regarded as *physical* processes subject to *mathematical* laws, the characterization of physical causes came to be seen as intrinsically mathematical in a way that outstripped the view of mathematics as merely being of instrumental use. A term that had come into use in the early seventeenth century to accommodate these trends, was 'physico-mathematics'. As Peter

[1] This point is forcefully argued in Shapiro (1974). I have benefited greatly from the account of the relationship between the hydrostatics of Descartes and Newton presented in that article. I complement it in later chapters by insisting on the importance of Boyle's theoretical innovations for an understanding of Newton. I also take issue with some details of Shapiro's interpretation of Newton's hydrostatics in Chap. 9.

Dear (1995, Chap. 6) has noted, that term was used by Jesuits early in the seventeenth century to accommodate the increased reliance on mathematics in the categorization of natural phenomena. I will use the term physico-mathematics in roughly this sense to refer to Descartes' adoption and extension of the traditional sciences of his day, especially optics and mechanics. As distinct from physico-mathematics in this sense, I refer to Descartes' attempts to reduce the phenomena to the results of the motions and arrangements of particles of matter/extension as his mechanical philosophy or simply as his philosophy. The following study will bring out the importance of this distinction, whatever words are used to describe it.[2]

Descartes was well up with and a capable contributor to the physico-mathematics of his day. This was evident, for example, in his work on optics, his deployment of the theory of simple machines (statics) together with his attempts to proceed beyond it to kinematics and dynamics, and his detailed knowledge and exploitation of Copernican astronomy. However, development of those sciences was not the main way in which Descartes was to transcend the neo-scholastic heritage passed on to him at *La Flèche*. In a way that came to fruition in the *Principles of Philosophy*, Descartes sought to replace the Aristotelian characterization of the fundamental ontology of the natural world as one of matter animated by forms. Instead he offered a world made up of nothing other than corpuscles of matter/extension each with some degree of motion or rest. Seeds of this conception seem to have been sown in Descartes' mind as a result of his early exchanges with Isaac Beeckman in 1618/19.[3]

Descartes' first attempt to give a systematic account of the mechanical worldview, aborted in 1633 in light of Galileo's condemnation, was *Le Monde*. This was eventually replaced by the more mature, and published, version articulated in *Principles of Philosophy.* In principle, Descartes' philosophy was to be capable of explaining or correcting physico-mathematics and, in that sense, rendering the latter redundant or as something less than genuine knowledge of the material world. In practice and to some extent covertly, the mature Descartes continued to contribute to developments at the physico-mathematical level. As a result there was an uneasy relationship between the philosophical and physico-mathematical endeavors. In his *Principles* and the formative versions that preceded it, Descartes drew on physico-mathematical results that he presented as analogies but which were crucial to Descartes' arguments in a way that transcends their categorization as merely analogical.

---

[2] In his study of Descartes, Schuster (2013, especially pp. 88–93) employs the term to describe both Descartes' reworking and extension of the mixed mathematical sciences and also to his various moves towards a corpuscular explanation of the phenomena involved. In this Schuster is adopting the practice of Descartes himself and of Beeckman, whom Descartes was following in that respect. I prefer to restrict the term physico-mathematics to a denotation of the various transformations of what Aristotelians had called mixed mathematics, in optics, mechanics and astronomy for example, and operated with concepts such as weight and light ray which were taken for granted and remained unexplained. In contrast to physico-mathematics was Descartes' grandiose philosophical project involving the attempt to reduce them to a mechanical system of matter in motion.

[3] For details see Schuster (2013, pp. 104–112).

In the following I discuss Descartes' engagement with hydrostatics at the level of physico-mathematics before broadening the discussion to the context of Descartes' mechanical philosophy. I begin with scrutiny of his early, unpublished, manuscript.

5.3 Descartes' Hydrostatic Manuscript

When Beeckman first brought Stevin's hydrostatics to Descartes' attention, in 1618, he was concerned to get a grasp of the mechanical causes responsible for the hydrostatic phenomena described by Stevin's theorems, something that was not supplied by Stevin's abstract 'proofs'.[4] Descartes' written response to Beeckman's queries have survived in what Stephen Gaukroger and John Schuster have described as 'the hydrostatic manuscript'.[5] In this section I present an interpretation of Descartes' position as presented there that is a little more charitable to Descartes than the one offered by Gaukroger and Schuster. I then follow my interpretation with a critical discussion of its significance.

Beeckman sought Descartes' views on the puzzling consequence of Stevin's theory that the force with which a body of water of given height weighs down on a surface which supports it is independent of the amount of water.[6] This can be seen by comparing situations B and D in Descartes' diagram, reproduced in Fig. 5.1. Here the areas of the bases, *gh* and *il*, are identical, as are the vertical heights, *f*B and *n*D. Beeckman urged Descartes to provide a causal account of why the force on the

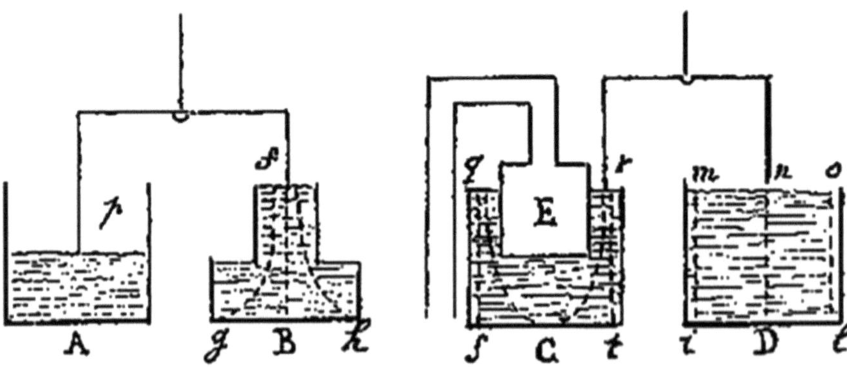

Fig. 5.1 The diagram in Descartes' hydrostatic manuscript (Descartes 1964–76, Vol. 10, p. 69)

[4] For an account of Beeckman's mechanical philosophy, and for his debt to Stevin, see Van Berkel (2013) especially pp. 130–136.

[5] See Gaukroger and Schuster (2002). I am grateful to John Schuster for making his translation of the manuscript from Latin available to me.

[6] Beeckman was concerned with the kinds of situation discussed by Stevin that I illustrated with my Fig. 1.1.

two bases, *gh* and *il*, are the same in spite of the fact that less water bears down on the former than on the latter.

In his response, Descartes noted that if the water in B and D were allowed to flow out of the respective containers at their bases, then the initial rate at which the water level would fall in B would exceed the rate at which it would fall in D. This follows from the difference in cross-sectional area of the two bodies of water in the upper regions once the continuity of the water is granted. The assumption of continuity, that the water descends 'regularly and in a mathematical manner' as Descartes expressed it, can be granted on empirical grounds, since no gaps appear in water steadily flowing out of a container. Building on the recognition of the unequal velocities, Descartes proposed that the degree to which a body weighs down depends not only on the amount of matter involved but also on the speed at which it would move in the first instant of its falling, 'for this speed contributes to the weight no less than the quantity of the body'. Given this, and given that the surface area at *f* is one third the surface area at *mno*, the water in the narrow, upper region, of B will weigh down three times *more* than the water covering the same height in D to the extent that it descends (or has the tendency to descend) three times faster. On the other hand the water in the upper regions of B will weigh three times *less* than the corresponding amount in the upper regions of D because its volume is three times less. Since the magnitude of these two differences, one in speed of descent and one in amount of water, both depend in the same way on the ratio of the cross-sectional areas of the water surfaces in B and D, they exactly balance each other. This provides an explanation of why the differing amounts of water in B and D bear down on the respective bases to the same degree.

While my reconstruction does, I believe, capture the crux of Descartes' argument, it cannot be left there because it leaves out an important dimension of his discussion. Whereas my interpretation is framed in terms of continuous bodies of water, Descartes talks of atoms or points that he presumes to constitute the bodies of water in the containers. The symbols *f, m, n* and *o* denote points or atoms in the water surfaces and the symbols *g, h, i* and *l* represent points or atoms at the bases. Here I entirely accept the views of Gaukroger and Schuster to the effect that here we find Descartes adopting Beeckman's corpuscular picture of the make-up of material reality, a view that persists throughout, and is central to, Descartes' later work. In the more mature work, particles or corpuscles of matter/extension take the place of what here Descartes describes as points or atoms. In the hydrostatic manuscript Descartes talks of point *f*, in diagram B, pressing down on points *g*, B and *h* in the base and of points *m, n* and *o* in diagram D pressing down on points *i*, D and *l*. Descartes sets out to show that the weights pressing down on the two bases, *gh* and *il*, are the same because each point in the equally sized bases are pressed to the same degree. To do this he must confront the difficulty that there are three times fewer points (or atoms) in the surface of B than in the three times larger surface of D. In Descartes' words, it must be shown 'that point *f* alone presses *g*, B and *h* with a force equal to that by which *m, n, o* press the other three *i*, D, *l*.'. Each point in the surface of D can be paired with the point in the base vertically below it. This unambiguous pairing cannot be done for B because there are three times fewer points in the top than in the bottom surface. Each point, such as *f*, must be paired with three points, such as *g*, B and *h*. Nevertheless, this sharing of the weighing down of point *f* over three points in the base is compensated for by the fact that *f* weighs down three times more than a point in the surface of D, such as *n*, because its tendency to move downwards, made manifest by the fact that the speed with which it does fall when allowed to, is three times greater. In this way Descartes concludes that the force on the bases in B and D are identical because the force on each point of each base, which are equal in area and so have the same number of points, are identical.

Gaukroger and Schuster criticize Descartes' account for a number of reasons. They find the mapping of point *f* onto points *g*, B and *h* to be totally *ad hoc.*

> Descartes makes no attempt to justify the threefold mapping from *f*. He merely slips it into the discussion as an 'example' and then proceeds to argue that *given the mapping*, *f* can indeed provide a three-fold force to *g*, B and *h* (p.562, italics in original).

These moves are described by our authors as 'ad hoc manipulations' which 'masquerade as clarifications, while in fact they condition a progressive lack of contact with the original aims of the problem' (p. 563) They also have problems with Descartes' assumption that point *f* can have a 'triple speed', towards *g*, B and *h*, given that it can have only one actual speed. (p. 564).

There is a charitable reading of Descartes' position that renders these criticisms misplaced. I suggest that the broken lines joining points in the surface to points in the base of containers in Descartes' diagrams represent flow lines, lines representing the path that points or atoms would follow were the containers allowed to empty via their bases. The flow lines also represent the line along which weight is transmitted at the first instant of motion. That is, they represent lines along which tendencies to motion act. Whatever the details of the flow, as a consequence of it each atom will follow one, unique, path from top to bottom. It remains to explain how there can be three paths emanating from the one point, *f*. There is a straightforward answer to this implicit in Descartes' text but admittedly not made explicit by him. Suppose a layer of water one atom thick were to flow out at the base of B. As a result the upper level will descend a space equal to three atoms. So the three atoms falling at point *f* will be available to eventually flow out at three points in the base. These latter points are labeled *g*, B and *h* in the diagram, but the exact point at which a particular atom starting from *f* emerges at the base need not be known. Whatever the details of the flow, each atom will follow a definitive path and the fact that the upper surface contains three times less points or atoms as the base is compensated for by the fact that points or atoms from the surface initially fall three times as fast, making three times as many atoms available. Beyond this, we do not know, and do not need to know, the details of the path that each atom takes. Nature can sort that out. The path of an atom need not even be continuous as it is jostled by adjoining atoms during its descent. Perhaps that is what Descartes had in mind when he explained that he 'drew the lines *fg*, *f*B, *mi* etc. not because we desired thereby a mathematical line of water would descend, but rather for easier comprehension of the demonstration'. Given this interpretation, the broken lines that Descartes offers 'for example' notionally represent the path of an atom and need not be specified further for the argument to go through. Descartes' position need not warrant the scorn directed at it by Gaukroger and Schuster.

My reading also shows the way towards Descartes' talk of tendencies to motion as distinct from actual motions. As Descartes declares, in relation to weights it is important to distinguish between 'inclination to motion and motion itself'. Heavy bodies have an inclination to move downwards but how they will actually move when free to do so will depend on the constraints to which they are subject. Because of the symmetry of the situation in D each atom will be free to follow its inclination to fall vertically downwards. This is not the case in situation B where the macroscopic continuity of the water can only be maintained if some atoms move around the corner to some degree as the water passes from the narrow to the broad part of the container. The motion of each atom will be constrained by those around it, with complicated paths resulting. 'For bodies that are inclined downwards are not inclined to move to the lower place with this or that speed, but rather they are inclined to move there as quickly as possible', that is, as quickly as the constraints permit. We have seen that for every atom falling in the broader regions of B there are three falling in the narrower regions. The paths of three atoms leaving the vicinity of *f* will diverge and end up at three different points on the base, which Descartes labels *g*, B and *h*, without thereby pretending that he can provide details of the motions that will determine which atoms end up where. It is in this sense that there can be an n-fold tendency near *f* leading to n distinct paths, one for each of n distinct atoms, where n is the ratio of the cross sectional area of the base to that of the upper surface of the water in B.

There remains a further clarification that needs to be brought to Descartes' exposition. The paradoxical situation in Stevin's hydrostatics that Descartes and Beeckman were seeking to illuminate involved differing amounts of water pressing equally on a given base, *but only on the assumption that those amounts of water possess the same height.* This dependence of the force on a base on the height of the water pressing on it was not made explicit by Descartes, but was implicit in his text. The dependence on height follows from the assumption that the pressing down of an atom is communicated to a base by way of its pressing on the atoms beneath it. As Descartes put it, we consider a point (atom) 'as it presses the rest of the bodies, not as it is impelled or attracted by others'. The contribution of each atom is cumulative and the total number involved will be proportional to the total height. This is how I read the significance of Descartes assertion that points on a base 'are each pressed by imaginable lines of water of the same length' where it is the perpendicular height to the surface that is the length in question. Of course, in summing the contribution of each atom, their differing perpendicular speeds must be taken into account from Descartes' point of view, with atoms in narrow regions pressing more than in wider regions of a container. It is in this way that Descartes accommodates the fact that the force per unit area on a base depends on the height of the body of water that rests on it, *and only on that height.*

The replacement of Gaukroger and Schuster's interpretation of Descartes' position by my more charitable one does not detract from a key point of theirs that is of important historical significance. The focus on inclinations to motion with magnitudes equal to that at the first instants of motions should they be permitted to occur was to become and remain a key feature of Descartes' mechanical system, especially the account of light central to it. The same can be said of the distinction between inclinations to motion and motion itself and the insistence that bodies can have only one actual motion but multiple tendencies to motion.

If I am right to interpret the dashed lines in Descartes' diagrams as representing flow-lines, then his account can be developed in a way that accommodates the isotropy of hydrostatic forces. In the situation depicted in situation B water will flow from holes in the side of the container and upwards from a hole in the upper surface at the level where the container broadens. The corresponding flow lines show the direction in which the water would flow were the corresponding holes opened. Consequently, these tendencies to motion can be interpreted as directions in which the water pushes. Once it is realized that water will leak upwards or sideways with a facility equal to that with which they leak vertically downwards through holes of equal cross section, one is well on the way to appreciating the complete isotropy of hydrostatic forces. Neither in his manuscript nor in his dealings with hydrostatics thereafter did Descartes move in the direction of acknowledging and accommodating such isotropy. When he was encouraged by Beeckman to respond to Stevin's hydrostatics, which, as we have seen, did accommodate isotropy, Descartes responded by considering the forces acting downwards on horizontal bases only. When, in his later work, he identified weight, and light, with centrifugal forces generated by aether vortices the treatment of hydrostatic forces as unidirectional was reinforced, as we will see.

5.4 Descartes' Adaptation of Archimedes

In this Section I gather together various passages in Descartes' work in which he has drawn on the Archimedean tradition in hydrostatics. Most of them occur in a context in which Descartes is developing details of his mechanical, corpuscular, system of the universe, but I defer consideration of that dimension of his work until later sections.

Descartes availed himself of the common-sense recognition that unlike a solid, neighboring parts of a fluid do not impede each other's motion.

> Of course, from the testimony of our senses, we recognize this difference to consist simply in the fact that the parts of fluid bodies easily move out of their places and consequently do not resist the movement of our hands into those places; while, on the contrary, the parts of solid bodies adhere to one another in such a way that, without sufficient force to overcome their cohesion, they cannot be separated. *Principles*, Part II, Section 54, Descartes (1983, pp. 60–70).

In *Le Monde* Descartes recognized that when a solid is immersed in a fluid less dense than it, it displaces a volume of fluid equal to its own with a consequent loss in apparent weight equal to the weight of fluid displaced.

> Thus they [the solid and the displaced fluid] are opposed to one another, each one being opposed to those that must replace it should it rise, and to those that must replace it should it descend, just as the two sides of a balance are opposed to one another (Gaukroger 1998, p. 49).

This is standard Archimedean talk with an explicit analogy with the balance added.

In the *Principles of Philosophy* (Part IV, Section 26) Descartes went beyond Archimedes by invoking the principle that 'no body can be carried downwards by its own weight, unless another body of equal size and less weight, is carried upwards at the same time'.[7] He deployed this principle in a way that led him to the conclusion that there are no internal pressures within a body of liquid. A drop of water surrounded by water 'is not pressed by the others' because 'if it were carried downwards, the other drops--- would have to rise in their place, which impedes their descent, since they are equally heavy'. That is, if a drop of water were to descend it would raise another drop equal in weight, thereby violating Descartes' principle. Therefore it does not descend and has no tendency to descend. This explicit denial of internal pressures by Descartes illustrates just how far from obvious was the construction of a conception of pressure suitable for a science of hydrostatics.

In the same context Descartes, like Archimedes, claimed that the base of a container of water is pressed 'only by as many drops as press it perpendicularly'. His example involved a container wider at the top than at the bottom and so his line of reasoning gave the right answer in this case, just as similar reasoning by Archimedes had led him to correct answers in simple cases of floating solids. But there are many instances where the assumption that it is only water perpendicularly above a base that contributes to the force on it gives the wrong answer. Situation B depicted in the diagram from Descartes' hydrostatic manuscript, shown in Fig. 5.1 is one of them.

Insofar as Descartes followed Archimedes in his hydrostatics, he inherited the limitations inherent in the latter's treatment of it. Because of the focus on weight and vertically acting upward thrusts that oppose it, Descartes did not acknowledge or accommodate the capacity of liquids to press in other directions. What is more, if the problem is to understand precisely how the weight of water acting downwards can be converted into an upward thrust, then Descartes' deployment of Archimedean considerations did not solve it.

Descartes went beyond considerations of hydrostatics at the physico-mathematical level to considerations that involved the corpuscles that he presumed liquids, solids, and indeed the entire Universe, to be composed of. There are two aspects of Descartes' deliberations that need to be appreciated in some detail in order to prepare for an examination of Descartes' hydrostatics at the corpuscular level. In the next Sect. 5.5, I outline Descartes' views on light and their development. The aim to give a mechanical account of light was a major factor guiding the construction of Descartes' mechanical philosophy and it led him to see light and hydrostatics as similar processes. In the following Sect. 5.6, I give an overview of

[7] A very similar principle appeared in a work on the motion of heavy bodies by Evangelista Torricelli (1644, p. 99) published in the same year as Descartes' *Principles of Philosophy.*

Descartes' mechanical system of the world and, in particular, its account of light and of weight. We will then be in a position, in subsequent sections, to pursue Descartes' hydrostatics at the corpuscular level and the analogies with light that were involved.

5.5 Descartes' Appropriation of Kepler's Optics

When Descartes turned his attention to optics around 1620 his starting point was the critique and extension of Witelo's optics in *Ad Vitellionem Paralipomena* by Johannes Kepler, published in 1604. This work represented the most up-to-date version of a tradition in optics that treated light in terms of linear rays and which dated back at least as far as Ptolemy. As Mark Smith (1987) has argued in detail, many of the ingredients that Descartes exploited on the way to the law of refraction were already present in the tradition that reached Kepler via Witelo. In that respect, Descartes' geometrical optics can be seen as a development of a practice dating back to Ptolemy just as his hydrostatics was part of a tradition dating back to Archimedes. However, Kepler's theory of vision marked a major break with the optical tradition insofar as he sought to further an understanding of the perception of light via the formation of images in the eye. Descartes took the break from the optical tradition further insofar as he sought to remove all traces of the Aristotelian notion that the physical properties of light are to be attributed to forms, in addition to matter, forms that are communicated to an observer in the act of perception. He sought a thoroughly mechanical account of the origins, transmission and perception of light.[8] As important background for an account of Descartes' geometrical optics I identify some basic assumptions informing the optics found in Kepler's writings that had origins in earlier work in the perspectivist tradition. Important amongst them were treatises by Alhazen in the eleventh century and Roger Bacon and Witelo in the twelfth century.

Light emanates as linear rays diverging in a spherically symmetric fashion in transparent media from all points in light sources, mainly the Sun and flames. Light intensity decreases with distance since rays are spread out over a greater area the greater the distance from the source. In the case of a spherical source such as the sun, rays perpendicular to the sun's surface are the strongest because the hemispheres of light emanating from each point on the spherical surface reinforce each other more in the case of rays directed radially than is the case for oblique rays. Rays falling normally on a surface illuminate it more intensely than oblique rays do because in the latter case a given parcel of incident rays is spread over a greater area. The transmission of light was typically, although not invariably, assumed to take place instantaneously. Nevertheless, instantaneous or not, the transmission of a ray is directed away from the source because it is in that direction that the intensity

[8] Mark Smith (2014) stresses the revolutionary aspects of the account of light given by Kepler and Descartes, thereby complementing the important continuity he had stressed in Smith (1987).

diminishes and also because it is the source that is the cause of the light that it transmitted from point to point through a luminous medium in a continuous chain from the source outwards.

Within the tradition I am describing, light was understood to be transmitted through transparent media more or less effectively in ways that are dependent on their density, and refraction was in some way connected with that assumption. From Alhazen onwards refraction was approached by splitting a ray incident on a boundary between two transparent media into normal and parallel components. The changes encountered by each component of the incident ray at the boundary were treated separately before recombining them into the refracted ray. Both the normal and parallel components were typically understood to undergo change and the details, on which there was no unanimity, were typically articulated with the help of mechanical analogies. For example, the dependence of the passage of light on the density of the medium was likened to the resistance experienced by a material object moving through a more or less dense medium, whilst Alhazen drew an analogy between refraction and the rebound of a projectile striking a thin plank obliquely. In the latter case the more normal is the line of impact the more likely the projectile is to break the plank, inviting a comparison with the fact that a light ray is the least diverted the closer it is to the normal.

Kepler's optical work drew heavily on the tradition I have described. He toyed with several ways of specifying the details of how the normal and parallel components of a ray of light might be affected at the boundary between two media and arrived at various trigonometric formulations of a law of refraction as a result. As Schuster (2013, pp. 153–163) has pointed out, when Descartes familiarized himself with thc optical tradition by studying Kepler's work he was particularly struck by one or two passages and diagrams in it which implied that light is transmitted with greater ease in a dense than in a less dense medium. This was an assumption that departed from the tradition and was at odds with what is suggested by the analogy with the resistance offered by a medium to the motion of a projectile.[9] In a manuscript fragment dating from around 1620 Descartes wrote:

> Because light can only be produced in matter, where there is more matter it is more easily generated; therefore it more easily penetrates a denser medium than a rarer one. Whence it happens that refraction occurs in the rarer medium from the perpendicular, in the denser medium towards the perpendicular. (Schuster 2013, p. 153, Descartes 1964–76, Vol. X, pp. 242–3).

When Descartes' words are matched to the passages in Kepler that probably inspired them, it is plausible to take them as implying that the ease of transmission of the normal component of a ray increases on passing into a denser medium. If that assumption is paired with the assumption that the parallel component remains unaffected then a law of tangents (rather than sines) governing refraction results. A law of tangents clashes with the empirical data assumed by Kepler which is perhaps why Kepler did not pursue the idea that the passage of light is facilitated by an

[9] See Schuster 2013, pp. 153–164 for details. My account of Descartes' debt to Kepler relies heavily on Schuster's analysis.

increase in density of the medium through which it passes. A few years later Descartes was involved in deliberations with Claude Mydorge about the construction of ray diagrams to represent refraction. In the course of them he replaced the assumption that the penetration of the normal component of a light ray is dependent on the density change at the boundary between two media by the assumption that the 'force' of a light ray, acting in the direction of the ray, is so dependent. That assumption, together with the assumption that the parallel component remains unaffected, led Descartes to the sine law of refraction.

In the perspectivist tradition the norm was to understand the transmission of light to take place instantaneously. Descartes was to adopt that assumption while giving it a mechanical interpretation. The idea that light is an instantaneously transmitted mechanical impulse had something in common with Descartes' conception of hydrostatic forces as characterized in his hydrostatic manuscript. The water in each part of the various containers considered by Descartes in the manuscript are assumed by him to bear down on the bases with a force proportional to the speed with which they would fall should they be allowed to do so. Each of the points or atoms making up the total mass instantaneously press down, combining to yield the net force on the base. It was not long before Descartes was drawing a direct comparison between the action of weight and the transmission of light. In a meeting with Beeckman in 1628 Descartes likened the refraction of light to the equilibrium of a balance with one of its weights immersed in a liquid. Descartes asked what angles the arms of an equal armed balance must be turned through to restore balance when one of the equal weights is immersed in a liquid. He concluded that balance occurs when the ratio of the weights is as the ratio of the sines of the angles between the balance arms and the vertical.[10]

Further to this, a comparison between the instantaneous transmission of light and the transmission of hydrostatic forces was explicitly invoked by Descartes in his *Optics*, written around 1630 and published along with his *Discourse on Method* in 1637. In the passage in question Descartes drew a comparison between the flow of wine from two holes in the base of a wine vat containing grapes as well as wine, on the one hand, and the passage of light from the sun to the two eyes of an observer via a transparent medium, on the other. Descartes assumes that light is transmitted from the sun via 'subtle matter'.

> Now, if you compare this subtle matter with the wine in the vat, and compare the less fluid or courser parts of the air and other transparent bodies with the bunches of grapes which are mixed in with the wine, you will readily understand the following. The parts of wine at one place tend to go down in straight line through one hole at the very instant it is opened, and at the same time through the other hole, while the parts at other places also tend at the same time to go through these two holes, without these actions being impeded by each other or by the resistance of the bunches of grapes in the vat. --- In the same way, all the parts of the subtle matter in contact with the side of the sun facing us end in a straight line towards our eyes in the very same instant they are opened, without these parts impeding each other, and even without their being impeded by the coarser parts of the transparent bodies which lie

[10] For a detailed analysis see Schuster (2013, pp. 199–203). To make this analogy work Descartes needed to assume that an object gains weight when immersed in a medium of greater density!

> between them. --- And note here that it is necessary to distinguish between the movement and the action or the tendency to move. For we may easily conceive that the parts of wine at one place should tend towards one hole and at the same time towards the other, even though they cannot actually move towards both holes at the same time, and that they should tend exactly in a straight line towards one and towards the other, even though they cannot move exactly in a straight line because of the bunches of grapes which are between them. In the same way, considering that the light of a luminous body must be regarded as being not so much its movement as its action, you must think of the rays of light as nothing other than the lines along which this action tends (Cottingham et al 1985, pp. 154–155).

Key notions, especially the distinction between motions and tendencies to motion, that were to become central to Descartes' mechanics generally and his account of light in particular, seem to have stemmed from his thoughts about hydrostatics, beginning with those evident in his hydrostatic manuscript of 1619.

As Schuster (2013, pp. 163–164) has stressed, in the 1620s Descartes set about transforming this unarticulated theory of light as an instantaneously transmitted mechanical impulse into a precise, corpuscular-mechanical theory, itself incorporated into a corpuscular-mechanical account of the workings of the Universe in its entirety. Here Descartes' efforts were conditioned by another branch of physico-mathematics with which he was thoroughly familiar, namely Copernican astronomy. The result was *Le Monde*, begun in 1629, abandoned in an unfinished form late in 1633 and unpublished in his lifetime. The central relevance of Descartes' conception of light to the mechanical synthesis described there is signaled by the fact that Descartes' subtitle for *Le Monde* was *A Treatise on Light*. Since Descartes did not address hydrostatics in a work designed for that purpose after the hydrostatic manuscript, his views on that topic need to be extracted largely from the details of *Le Monde* and the more mature and considered version of Descartes system published as *The Principles of Philosophy* in 1644, as well as some of his correspondence.

5.6 Descartes' Mechanical Universe and His Treatment of Weight and Light

I make no attempt here to engage with the details of the mechanical picture of the world sketched in *Le Monde* and articulated in the *Principles of Philosophy.* I will simply summarize those of its central features sufficient for my purpose.

Descartes' material Universe is made up of nothing other than particles of matter, characterized by their extension, individuated by their location and possessing some degree of motion or rest. The intimate connection between matter and extension has the consequence that the Universe is completely full of matter, with no void spaces. Since it is a plenum, all motions in the Universe must involve a circulation of matter in closed circuits. The particles making up the Universe are divided into three classes or 'elements'. The Sun, and fixed stars which shine with a light of their own, are composed of particles of first element that are small and fast-moving. The space separating the stars, planets and comets is occupied by spherical particles of second element. Planets, including the Earth, and comets are made up of relatively large particles either at rest with respect to each other, as in solids, or moving with speeds relatively slow compared with those composed of first and second element, as in fluids. Spaces between particles of second and third element are filled by fine particles of first element, ensuring that the Universe is entirely full. The three elements map onto three basic properties of light, the rapid motions of first element in the sun and stars giving rise to light, the spherical particles of second element transmitting it and the particles of third element constituting the earth, planets and comets reflecting, refracting or absorbing it.

The particles of matter in the Universe as characterized above are governed by 'laws of nature'. The first of them requires that the total force of motion in the Universe be conserved, where that total force is the sum over all particles of the product of volume and speed for each of them. The results of impacts between particles are governed by laws of collision, which collectively correspond to Descartes' third law of nature.[11] The second law is of central importance for our focus on hydrostatics. In the *Principles* it reads, 'all movement is, of itself, along straight lines; and consequently, bodies which are moving in a circle always tend to move away from the center of the circle which they are describing' (*Principles*, II, 39, Descartes 1983, p. 60). Descartes makes it clear that the motion in straight lines is an instantaneous tendency that will rarely lead to actual motion in a straight line because of the influence of particles surrounding a body that tends to move. A body moving with a circular motion, in addition to the tendency to move in a straight line tangent to the circle, will also have a tendency to move away from the center since the tendency to linear motion possessed at each instant would increase the distance of the body from the center should it eventuate. Descartes illustrated this point by reference to the circular motion executed by a body whirled on a sling. Here the instantaneous tendency of the body to move in a straight line tangent to the circle is evident from the initial motion of the body if it is freed from the sling and the tendency to move away from the center can be felt in the hand that whirls the sling. Instantaneous tendencies to motion, which had previously figured in Descartes' hydrostatic manuscript, here take a prominent place in the very formulation of the laws of nature that govern the mechanical Universe and prior to his application of them to specific phenomena.

With this corpuscular picture of the Universe and the laws of nature governing it assumed, Descartes aimed to explain broad features of 'the Visible Universe'. Among the phenomena to be explained were our planetary system understood in broadly Copernican terms, the generation of light in the Sun and fixed stars and its transmission directly to earth and also indirectly by reflection from planets and

[11] Descartes' analysis of collisions involved 'determinations' of a motion, which were directed, vectorial quantities, unlike the quantity of motion figuring in the first law. I do not pursue the details here. As we shall see, Descartes' hydrostatics was entirely qualitative and the difficulties associated with it were not of a kind that could be dealt with by precise mathematical formulations. A charitable rendering of Descartes' notion of determination and his application of it to the laws of motion can be found in McLaughlin (2000) and in Damerow et al. (2004, pp. 94–133).

comets, the weight displayed by terrestrial objects and the nature and motions of comets and sunspots. The cosmology invoked by Descartes involved vortices composed of second element circulating around the Sun and fixed stars. These vortices carry the planets with them and trap them in stable orbits and are responsible for the less regular motion of comets from the vortex of one 'sun' to that of another. Secondary vortices of second element circulating the Earth and other planets serve to explain weight and the motion of moons around their parent planets. In the main, Descartes' Universe was in a steady state, with planets in each 'world' in stable orbits around their sun and with the multiple worlds in dynamic equilibrium with each other. Exceptions were the motions of comets and the fall of terrestrial objects. With this sketch of Descartes' mechanical conception of the Universe in place, I focus on two issues that were particularly relevant to hydrostatics, namely, his account of weight and his account of light rays.

Portions of second element constituting the secondary vortex orbiting the earth will possess a tendency to move away from the center of rotation as specified by the second law. This tendency will not typically result in an actual motion of a portion of second element because it will be surrounded and inhibited by similar portions of second element with a similar tendency to move outwards. This equilibrium situation is upset by the presence of a terrestrial object since such an object will not share the circulatory motion of the vortex, which is distinct from and faster than the rotation of the Earth, and so will lack its tendency to move outwards. The deficiency of outwards pressing brought about by the presence of the terrestrial object is felt by the latter as an inwards force towards the center, a conclusion helped by the assumption that a void is impossible. As a result, a terrestrial object free to fall will do so. As Descartes explains 'the weight of each terrestrial object is not, strictly speaking, produced by all the heavenly matter flowing around it, but rather only by that portion of the heavenly matter which immediately ascends into the place of the descending body, and which, therefore is exactly equal to it in size' (*Principles*, IV, 23, Descartes 1983, p. 191). The weight of a body is determined by the amount of third element it contains.

The tendency for matter moving in circles to move away from the center of rotation was as central to Descartes' account of the generation of light and its transmission from Sun to Earth as it was to his account of weight. The vortex of second element around the Sun that carries the planets and earth with it is in a steady state, with each spherical particle of second element experiencing a tendency to recede from the center while constrained by neighboring particles with a similar tendency. If we consider a tube of second element stretching from the Sun to the eye of an observer on earth, then a cumulative pressing of sphere upon sphere emanating from the Sun will be instantaneously transmitted to the eye. This pressing constitutes seeing, and the line of transmission of the pressing constitutes a ray of light.

As we have seen, the fact that light extends in straight lines was an assumption that Descartes shared with the majority of contributors to the optical tradition from Ptolemy to Kepler. That assumption could reasonably be taken as a brute phenomenological fact borne out by such things as shadows and the way in which a beam is transformed into a ray on passing through a small aperture. But neither Descartes nor the scholastic perspectivists on whom he drew and aimed to transcend were prepared to leave it at that. The latter set their account in a philosophical framework involving an appeal to forms as well as to matter. It was the form characteristic of luminous objects that gave them the power to generate luminosity in neighboring media and it was the form characteristic of transparent media that led them to respond to this power by becoming luminous. Descartes reacted against such an approach in a way that was to become typical within the 'mechanical philosophy' that he was setting in train. He had already indicated the problems he had with the ontological status of 'those little images flitting through the air, called intentional forms' in his *Optics* and prepared the way for their elimination by comparing the act of seeing with the way in which a blind man connects with the world by probing with a stick.

> For, just as our blind man can feel the bodies around him not only through the action of these bodies when they move against his stick, but also through the action of his hand when they do nothing but resist the stick, so we must acknowledge that the object of sight can be perceived not only by means of the action in them which is directed towards our eyes, but also by the action in our eyes which is directed towards them (Cottingham et al 1985, p. 154).

Insofar as the latter actions are 'nothing other than light' the phenomenon is explained mechanically and obviates the need to appeal to forms, transmitted through transparent media and apprehended by an observer. Given that this was Descartes' objective, it was important for him to spell out the details of the mechanisms involved in optical phenomena. Only by doing so could he free himself of analogues of the charge leveled at the scholastics by the mechanical philosophers, namely, that their appeal to forms as explanatory lacked adequate specification, were merely verbal, or were circular. Given the nature of his project, Descartes needed to specify the mechanisms responsible for light and for material phenomena generally, and strived to do so in terms of corpuscle pushing against corpuscle.

The pressing transmitted via light rays and the pressing due to terrestrial weight are alike in that they both have their root cause in the centrifugal tendency of circulating spheres of second element, and, as such, act linearly. It is therefore not surprising to find Descartes illustrating his account of light as impulses instantaneously transmitted via an expanse of spheres of second element using analogies involving the weight of adjacent corpuscles resembling billiard balls. I focus on some details of the way in which he did so in the following two sections.

5.7 Descartes' Distinction Between Solids and Liquids at the Corpuscular Level

The nature of fluids as opposed to solids was central to Descartes hydrostatics and his theory of light. We saw in Sect. 4.4 that Descartes adopted the common recognition that neighboring parts of a liquid are not constrained by each other in the way that parts of a solid are. Not content with this, in *Le Monde* and in the *Principles of Philosophy* he theorized about the distinction between solids and liquids at the corpuscular level. He held that the corpuscles of solids cohere in a way that they do not in liquids. According to Descartes, from the common recognition mentioned above 'it may be concluded that those bodies which are divided into very small parts which are agitated by a diversity of movements are fluid; while those bodies whose particles are all contiguous and at rest, are solid' (*Principles*, II, 54, Descartes 1983, p. 70).

The understanding that the particles of a solid are contiguous and *at rest* was of special significance for Descartes. For him rest was not a special case of motion, namely zero motion, but was distinct from motion and logically contrary to it.[12] So the force necessary to separate contiguous particles at rest with respect to each other is very large compared to that required to separate particles of a fluid which are already moving. For Descartes it is not simply the case that adjacent parts of a liquid offer little resistance to being moved. The particles composing a liquid are constantly in motion, and this is why adjacent parts of a liquid are easily moved relative to each other in a way they are not in a solid, whose parts are at relative rest, the logical contrary of motion.

The explanation of solidity by appeal to rest was important for Descartes' project insofar as it involved reducing the material world to matter and motion and nothing else. Insofar as the difference between solids and liquids could be explained by appeal to motion and rest, no entities as ontologically suspect as Aristotelian forms needed to be invoked. Descartes made the point explicit. Commenting in the *Principles* on what it is that holds the parts of a solid together Descartes writes as follows:

> Furthermore, our reason certainly cannot discover any bond which could join the particles of solid bodies more firmly together than does their own rest. For what could this bond be? It could not be substance, because there is no reason why these particles, which are substances, should be joined by any substance other than themselves. Nor is it a mode different from rest; for no other mode can be opposed to the movement which would separate these particles more than is their own rest. Yet, besides substances and their modes, we know no other kinds of things. (*Principles*, II, 55, Descartes 1983, p. 70)

In a similar context in *Le Monde* Descartes wrote:

> Thus, to make the hardest body imaginable, I think it would be enough for all the parts to touch each other, with no space remaining between any two and none of them in the process of moving. For what glue or cement can one imagine beyond this with which to hold the one to the other?'[13]

---

[12] The details of Descartes' adoption of a logic of contraries are identified and spelt out in Damerow et al. (2004, pp. 81–94).

[13] Gaukroger (1998, p. 10). Descartes' distinction between solids and liquids by appeal to rest as the 'glue' that holds together the parts of the former fails to capture the common sense fact that, whilst its shape can vary, the size of a portion of liquid remains constant.

For Descartes the only material substance is matter/extension and its only modes are rest or motion and its determinations. Nothing further, akin to Aristotelian forms, is required.

Descartes' elaboration of the distinction between solids and liquids at the corpuscular level was of key importance from the point of view of his mechanical philosophy with its key aim to eliminate Aristotelian forms and reduce the world to a system of corpuscles possessing only shape, size and a degree of motion or rest. However, if we abstract from Descartes' ambitious project and retain our narrow focus on the development of hydrostatics, it is doubtful whether the details of Descartes' distinction between solids and liquids can be said to have played a constructive role.

5.8 Corpuscular Models and Isotropy

We have seen that in Descartes' mechanical system the behavior of light and of liquids could be traced to similar causes, the centrifugal forces generated by vortices being responsible both for light rays from the sun and the weight of liquids. In Descartes' full universe both light and hydrostatic forces involved the instantaneous transmission of tendencies to motion that are linear. Given this it is not surprising to find Descartes modeling optical phenomena by systems of corpuscles pressing down by virtue of their weight. The particles of second element responsible for the transmission of light from the sun were spheres according to Descartes, and their radius increased with distance from the sun, a fact that was crucial for Descartes' explanation of how planets get trapped in stable orbits. As a consequence, neighboring spheres are of the same size, so it was appropriate for Descartes' corpuscular models to involve uniformly sized spheres akin to billiard balls.

As I have stressed, the transmission of light and hydrostatic forces both stem from the centrifugal forces generated by vortices in Descartes' system, so they are essentially linear. There were problems lurking here, some internal to the details of Descartes' corpuscular models and others involving a clash with departures from linearity exhibited by the behavior of liquids.

If we take the static arrays of corpuscles frequently appealed to by Descartes in *Le Monde, The Principles of Philosophy* and his *Optics* at face value then they pose two kinds of problem. One is that the forces transmitted by way of corpuscle pushing against corpuscle depart from linearity except in the special case where their centers are aligned in the direction of the pushing. The details of the way in which forces are transmitted depend on the details of the corpuscular arrangements. Another difficulty is that systems of corpuscles, representing a liquid in a container, say, can become locked. Two billiard balls touching each other at a point above a hole that is a little larger than they are will not fall through the hole.

Descartes recognized both difficulties and responded to them in the *Principles of Philosophy.* It is one of the rare instances that Descartes actually used his assumption that the particles making up a fluid are in constant motion. Because of this, says

Descartes, the arrangements of corpuscles depicted in his diagrams are snap shots of situations that are very short-lived. Descartes distinguished between the coordinated movements of corpuscles such as those shared by particles of second element rotating in a vortex or by particles of a bulk of liquid participating in some motion as a whole, on the one hand, and the uncoordinated, agitated motions, on the other. Corpuscles of second element will push their neighbors now a little to the left, now to the right but these movements will cancel out on average leaving the general circulation of the vortex and the consequent tendency for the corpuscles to radiate outwards. In a similar kind of way a liquid in a vessel will have a general tendency to move downward in spite of the agitations in other directions. Arrangements that threaten to lock fluids into a fixed array will only be transitory. (The billiard balls locked above a hole in my example can be made to fall through the hole if the container is shaken.) Here are Descartes' own words:

> But the same thing is not true of the little globes which form the second element; for although it sometimes happens that they are arranged like those shown in this figure, they however only remain in this position for an instant, because they are perpetually in motion and so continue to move without interruption. Moreover, it must be noticed that the force of light does not consist in duration of motion, but only in the pressing or first duration of motion, even though actual motion may not result from this process (*Principles,* III, 63, Descartes 1983, p. 117).

> Finally, it must be added that, although the particles of heavenly matter are agitated at the same time by many diverse movements, yet all of their action harmonizes and, as it were, counterbalances one another in such a way that, due solely to their encounter with the bulk of the earth which resists their movements, they strive to move away equally, as if from its center (*Principles*, IV, 27, Descartes 1983, p. 193).

I have pinpointed here ways in which Descartes responded to the threat that corpuscles interacting in ways other than along their line of centers posed for his assumption that light and weight acts linearly. But this solution led to a serious conundrum for him at a point where he needed some departure from linearity. This point concerned the appearance of the Sun as a disc in the sky. Given that rays of light represent centrifugal forces directed away from the axis of rotation of the Sun the latter should appear as a point when viewed from the Earth, situated on the ecliptic. The perspectivist tradition had not been faced with this difficulty because it included the assumption that the sun radiates from all points on its surface and not only via rays emanating from the center. Descartes tackled the problem by invoking the ways in which the interactions of adjacent corpuscles can diverge from linearity. But he had to carefully manage this conception to accommodate the fact that the Sun appears as a small disc rather than a greater expanse. Schuster (2013, pp. 501–509) and Alan Shapiro (1974, pp. 254–260) have subjected Descartes' argument to critical scrutiny and have shown how he delimited the way in which the mutual pressings of spheres can reach us from the Sun in an ad hoc way to give him the result that he wanted. It should also be noted that, since the argument invokes specific arrangements of corpuscles, it runs counter to Descartes distinction, invoked elsewhere as we have seen, between the general motions and tendencies to motion resulting from vortex motions and the irregular motions of corpuscles on which they are superimposed.

Insofar as Descartes' theory of light and his hydrostatics involved linearly directed motions or tendencies to motion it failed to come to terms with a range of phenomena that run counter to that assumption. A departure from strict linearity was already in evidence in Descartes' hydrostatic manuscript. For, although his emphasis was on the liquid pressing down on the base of the various containers shown in his diagrams, the curved lines shown in Figs. B and C in Fig. 5.1 indicate the curved paths that portions of liquid would take were liquid to drain out of the base of the containers B and C. Had Descartes followed this through with the recognition that liquid will flow horizontally and vertically from suitably situated holes in the containers he would have been on the path to the recognition of the isotropy of hydrostatic forces. While Descartes may not have had the relevant phenomena in mind at this stage, the same cannot be said of a mature Descartes who actually discussed experiments on the horizontal efflux of liquids from containers in a letter to Huygens in the context of the law of fall.[14]

A similar issue arises in the context of Descartes' analogy, in his *Optics*, between wine flowing from a vat partly filled with grapes and light passing through a transparent body, discussed in Sect. 5.5. Here the wine is likened to the second element through which light is transmitted and the grapes are analogous to the particles of third element that constitute the transparent body. In the case of light passing through a transparent body it is Descartes' view that there is no actual motion but only a tendency to motion. But in the case of the wine there is a motion and it departs from linearity insofar as the wine needs to flow around the grapes. The problem for Descartes is compounded once it is acknowledged that wine will freely flow from a faucet in the side of the vat.

A number of the phenomena described in or implied by Stevin's *Elements of Hydrostatics* spell serious trouble for Descartes position on hydrostatics and light. The situation depicted in Fig. 5.2 is not one that to my knowledge was invoked or discussed in Descartes' time but the phenomenon that it illustrates is a consequence of Stevin's law. The force on the container shown will be equal to the weight of a cylinder of water with a cross section equal to that of the base and a height equal to the total height of the water. Here the pressing of the water diverges through the aperture, A, in the dividing diaphragm so that the resultant pressing on the base is as if that diaphragm did not exist. This is a threat to the corpuscular explanation generally since one would expect the diaphragm to support the corpuscles above it, so lessening the force on the base. For Descartes, insofar as the pressing down of a terrestrial fluid and the pressing that is light are similar processes, my example undermines the assumption that light travels in straight lines. If the centrifugal tendency underlying the weight of water diverges in passing through an aperture, so should the centrifugal tendency constituting light. Here my position is totally in accord with Shapiro's insistence that the linearity of light rays was a central assumption of Descartes' mechanical system of the world to which all other details were

[14] Descartes (1964–76, III, pp. 807–808), letter to Huygens, Feb. 18, 1643.

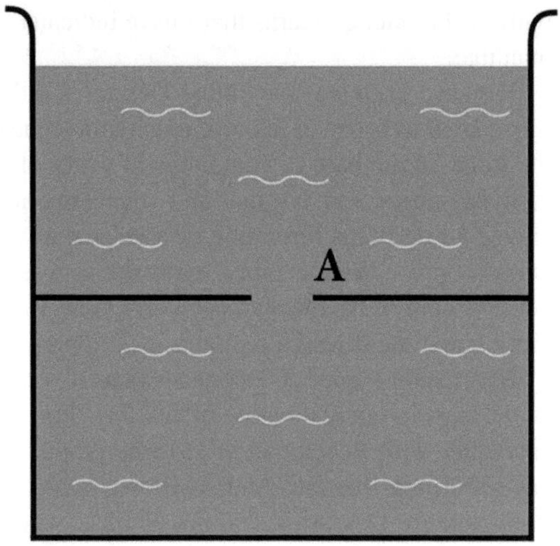

Fig. 5.2 The pressure of water on the base of the container is unaffected by the presence of the diaphragm containing aperture A

subservient.[15] Once Descartes had identified light rays with linear centrifugal tendencies he could not afford to acknowledge isotropy as exhibited by fluids. As Shapiro (1974, p. 254) puts it,

> Although Descartes' concept of centrifugal *conatus* implies a proper concept of pressure, he could not fully pursue this implication – as Newton did – for to do so would violate the law of rectilinear propagation for light. On this last point Descartes was as firm as his most severe critic, for he unquestionably considered light to consist of rays extending in straight lines. Descartes could take a small number of steps towards admitting a pressure, but these were concessions he had to make if his theory was to agree with the observed phenomena [such as the appearance of the sun as a disc]. Only so far would he go, but no further.

5.9 An Assessment of Descartes' Dealings with Hydrostatics

My focus in this book is a narrow one, the emergence of a science of hydrostatics in the seventeenth century. There could hardly be a bigger contrast between the narrowness of that concern and the broad concerns with which Descartes struggled. His attempt to construct a mechanical, corpuscular account of the workings of the world in its entirety was a grand enough project to be sure, but it formed only a part of what the mature Descartes aspired to provide. Among his other ambitious projects were the establishment of a dualist metaphysics, construction of a universal method, innovations in mathematics sufficient to rank him amongst the most able mathematicians of his time and a reconstruction of the philosophy of perception. If Descartes' contributions to hydrostatics were insignificant, as I believe they were, then there might well be an inclination to excuse him on the grounds that he was preoccupied

[15] This is a key theme of Shapiro (1974).

with the broader concerns that I have indicated. However, there is a little more to it than that.

We have seen that the young Descartes did take an interest in hydrostatics and formulated a theory of his own to explain some counter-intuitive claims concerning the force on the base of containers of water of various configurations. This version of hydrostatics was original and suggestive and, what is more, it seems to have played a role in his formation of a science to which he did significantly contribute, namely optics. What is more, hydrostatics was involved in fundamental ways in the development of his mechanical world view, in part because it became necessary for him to treat the celestial region as a *fluid* plenum.

Hydrostatics aside, Descartes was well versed in and able to make contributions to the physico-mathematics of his day. For instance, from the time of his early encounter with Beeckman in 1618 he grappled with the problem of formulating a law governing free fall. And, as shown by Peter Damerow et al. (2004, Chap. 1), it was a mature Descartes, and not Galileo, who, in 1643, first explicitly formulated that law as the proportionality between velocity and time of fall in the modern way, although without appreciating the full implications of his achievement.[16] Notwithstanding the fact that Descartes embedded his account of light in his mechanical philosophy in a way that was to prove incorrect, he did make significant contributions to geometrical optics by formulating the law of refraction and employing it, for example, in lens theory and in an explanation of the rainbow. His detailed engagements with astronomical phenomena such as comets and sunspots show the extent to which his knowledge of Copernican astronomy and the relevant empirical data were well up to date.

As can be deduced from Descartes' letter to Christian Huygens on free fall in 1643, his interest in and contributions to physico-mathematics continued into his maturity.[17] However, from the time he took on the task of constructing a mechanical account of the workings of the world in its entirety in the early 1630s, Descartes refrained from making public his continuing explorations at the physico-mathematical level. From the point of view elaborated in the *Principles of Philosophy,* laws at the latter level were either to be explained as consequences of his mechanical theory, such as he believed he had done in the case of the law of refraction, or dismissed as mere approximations that could depart from the truth as specified in his mechanical theory, as he believed to be the case with the law of fall. Daniel Garber (2000, pp. 113–130) has explored the contrast between the mature Descartes' private engagement with physico-mathematics, evident from his correspondence, and his public stance according to which such endeavors were misplaced or redundant in the light of his mechanical philosophy.

[16] Details can be found in Damerow et al. (2004, Chap. 1). An English translation of the relevant section of Descartes' letter to Huygens in 1643, in which the law of fall was expressed, is reproduced in Damerow et al. (2004, pp. 280–281). See also Schuster (2013, pp. 128–153).

[17] See Damerow et al. (2004, pp. 280–281) for a discussion of Descartes' letter and its implications.

Given Descartes' continuing engagement with physico-mathematics, the quality of many of his contributions to it and the extent to which hydrostatics impinged on his concerns, some diagnosis of the shortcomings of his contributions to that science would seem to be in order. They can be traced to Descartes' ambition to reduce the Universe to a corpuscular, mechanical system and the degree to which, from the beginning, that endeavor was conditioned by a particular mechanical conception of light as a mechanical impulse transmitted instantaneously through the heavens. According to Descartes, liquids press down by virtue of their weight and light presses on the eye by way of an impulse transmitted through the heavens from the sun, both having their root cause in the centrifugal forces generated by vortices of heavenly matter. Given such a conception Descartes could not, or could not afford to, acknowledge the isotropy of hydrostatic forces. As I have stressed, this isotropy was already included, although not adequately theorized, in Stevin's *Elements of Hydrostatics* and was signaled by common phenomena known to Descartes.[18] A property of fluids is that they convert directed forces applied to them into isotropic ones. To admit this would have forced Descartes to give up his commitment to the fact that light extends in straight lines. In this respect I am merely reiterating the point stressed by Shapiro (1974, especially p. 252).

Descartes did not make significant moves towards the technical conception of pressure that was to prove necessary for an adequate theorization of hydrostatics. That move proved to be far from obvious, and it was 68 years after the publication of Stevin's *Elements of Hydrostatics* before Blaise Pascal made the first significant steps in that direction. That is the topic of the following chapter.

References

Cottingham, J., R. Stoothoff, and D. Murdoch. 1985. *The philosophical writings of Descartes*. Vol. 1. Cambridge: Cambridge University Press.

Damerow, P., G. Freudenthal, P. Mclaughlin, and J. Renn. 2004. *Exploring the limits of pre-classical mechanics: A study of conceptual developments in early modern science*. New York: Springer.

Dear, P. 1995. *Discipline and experience: The mathematical way in the Scientific Revolution*. Chicago: University of Chicago Press.

Descartes, R. 1964–76. *Ouevres de Descartes* Vol. 12, eds. C. Adams and P. Tannery. Paris: Vrin.

———. 1983. *René Descartes: Principles of philosophy*. Trans. V.R. Miller and R.P. Miller. Dordrecht: Reidel Publishing Company.

[18]As Shapiro (1974, 252–253) has noted, there is an example of isotropy exhibited by air in *Principles of Philosophy*. In occurs in a passage in which Descartes attempts to reconcile the spherical symmetry of the sun with the cylindrical symmetry of the vortex of second element that circulates it. 'In the same way that we see a glass bottle become spherical solely from the fact that air is blown through an iron tube into its molten matter: because of course this air does not move from the mouth of the bottle towards its base with more force than that with which it is deflected from there to all other parts, and drives them all equally easily.' *Principles* III, 75, Descartes (1983, p. 125).

Garber, D. 2000. A different Descartes: Descartes and the programme for a mathematical physics in his correspondence. In *Descartes' natural philosophy*, ed. S. Gaukroger, J. Schuster, and J. Sutton, 113–130. London: Routledge.

Gaukroger, S. 1998. *René Descartes: The World and other writings*. Cambridge: Cambridge University Press.

Gaukroger, S., and J. Schuster. 2002. The hydrostatic paradox and the origins of Cartesian dynamics. *Studies in History and Philosophy of Science.* 33: 535–572.

Mclaughlin, P. 2000. Force, determination and impact. In *Descartes' natural philosophy*, ed. S. Gaukroger, J. Schuster, and J. Sutton, 81–112. London: Routledge.

Schuster, J. 2013. *Descartes-Agonistes: Physico-mathematics, method and corpuscular mechanism, 1618–33*. Dordrecht: Springer.

Shapiro, A. 1974. Light, pressure and rectilinear propagation: Descartes' celestial optics and Newton's hydrostatics. *Studies in History and Philosophy of Science.* 5: 239–296.

Smith, A. Mark. 1987. Descartes' theory of light and refraction: A discourse on method, *Transactions of the American Philosophical Society*.77: Part 3.

———. 2014. *From sight to light: The passage from ancient to modern optics*. Chicago: University of Chicago Press.

Torricelli, E. 1644. *Opera geometrica*. Florence: Amatoris Massae and Lauenti de Landis.

Van Berkel, K. 2013. *Isaac Beeckman on matter and motion*. Baltimore: Johns Hopkins University Press.

Chapter 6
Pascal's *Equilibrium of Liquids*

Abstract Pascal began his *Equilibrium of Liquids*, written around 1654, by noting some counter-intuitive experimental results. He then utilized what is now known as the hydraulic press to introduce his theory of how a liquid can function as a machine for multiplying force. A pressing at one location on the boundary of a liquid, measured as a force per unit area, is instantaneously transmitted to any other location on the boundary, however orientated, as the same force per unit area. This takes place on account of the continuity and fluidity of the liquid. In the remainder of his short book Pascal showed how his theory could account for a range of phenomena, some long familiar, such as floatation, and others less so, such as the deflation of an inflated bladder when lowered in water. The reliance on experiment stood in contrast to the demand that propositions of a science should be derived from postulates grantable at the outset. The significant moves made by Pascal towards the technical sense of pressure retained the common conception that pressure acts on the boundary between a liquid and a solid and did not consider pressures acting in the body of liquids. Pascal's, somewhat exaggerated, stress on the debt of his hydrostatics to novel experiments probably stemmed from his involvement with pneumatics.

6.1 The Context of Pascal's Hydrostatics

Following the publication of Simon Stevin's *Elements of Hydrostatics*, 68 years elapsed before a second treatise devoted to that topic was compiled. That work, *On the Equilibrium of Liquids*, was written by Blaise Pascal around 1654 but not published until 1663, a year after his death. The main purpose of this chapter is to analyze and appraise Pascal's hydrostatics as it is presented there with particular emphasis on how he moved beyond Stevin's *Elements* and how his approach differed in style from the Euclidean ideal that the latter aspired to follow. In this section I set some context for the discussion.

From early in his youth Pascal was participating in discussions orchestrated by Marin Mersenne in Paris, to which he was introduced by his father, Etienne Pascal, himself an accomplished mathematician. The Mersenne circle included Gilles Personne de Roberval, Girard Desargues and Pierre de Fermat, who were involved in contemporary developments in mathematics and experimentation. The circle also

A.F. Chalmers, *One Hundred Years of Pressure*, Archimedes 51,
DOI 10.1007/978-3-319-56529-3_6

maintained contacts with philosophers with more general concerns, including Rene Descartes, Pierre Gassendi and Thomas Hobbes. In the wake of Evangelista Torricelli's experiment in 1644 Pascal became involved in experiments of his own. He participated actively in debates about the significance of the space above the mercury in what we now would call the mercury barometer. Pascal sided with those who saw in this phenomenon an opportunity to strike a decisive blow against the Aristotelian world view, which they interpreted as ruling out the possibility of a void and attributing a range of effects, such as those exhibited by syphons and lift pumps, to nature's abhorrence of a vacuum.

Towards the end of 1654 Pascal had a religious experience that effectively put an end to his contributions to the new sciences. He became deeply involved in a highly rigorous and somewhat extreme form of Catholicism associated with the convent Port Royal and inspired by the writings of Augustine. His subsequent writings were indicative of this new involvement. However, Pascal's religious transformation did not take place before he had written two treatises, *On the Equilibrium of Fluids*, mentioned above, and a companion piece *On the Weight of the Mass of the Air*. These works were published, posthumously, by Pascal's brother-in-law Florin Perrier, in 1663, 2 years after the former's death. The remainder of this chapter is concerned mainly with the contents and style of *The Equilibrium of Liquids.*

6.2 Pascal's Style of Reasoning Compared to That of Stevin

Pascal made no explicit reference to Stevin's work in his treatises on liquids and on air, nor elsewhere as far as I am aware. However, there is no doubt that Pascal was familiar with it and owed a significant debt to it.[1] Many of the experiments that Pascal appealed to in the articulation of his hydrostatics are versions of the ones cited by Stevin as 'practical applications' of his own theory. Even in the unlikely event that Pascal had not consulted Stevin's text directly he is likely to have encountered the commentary on the latter in Mersenne's *Cogitata physico-mathematica,* published in 1644 and with which we know Pascal to have been familiar.[2]

There are qualitative differences between the styles of reasoning in the treatises of Stevin and Pascal that I wish to emphasize in the following, although I will subsequently need to qualify my contrast to a small degree. Unlike Stevin, Pascal was not so much concerned to derive theorems or propositions true of hydrostatics, but rather, was intent on understanding hydrostatic phenomena by identifying their causes. He sought to justify his theory, not by deriving it from evident postulates granted at the outset, but by appealing to the results of experiment. Whereas Stevin presented his hydrostatics as a mathematical theory conforming to the Euclidean

[1] The original, Dutch version of Stevin's *Elements of Hydrostatics* had been translated into Latin in 1608 and into French in 1634.

[2] Pierre Duhem (1905, pp. 602–604) made a case for Pascal's familiarity with the details of Stevin's *Elements of Hydrostatics.*

ideal, Pascal presented his as a physical theory supported by experiment. I elaborate on this theme in the remainder of this section.

The subtitle to Pascal's two treatises, on liquids and on air, reads as follows:

> Containing the explanation of the causes of various effects of nature which had not been known hitherto, and in particular of those which had been ascribed to the abhorrence of a vacuum.

Pascal here announces his intent to identify the causes of various effects. As far as *The Equilibrium of Liquids* is concerned we will see that those causes were located not only in the weight of liquids but in their 'continuity and fluidity'. Pascal's elaboration of fluidity took him some way towards the identification of the modern concept of pressure as the key cause characteristic of hydrostatic effects. There is a direct contrast with Stevin here. We saw in Chap. 3 that, even if we concede to Stevin the validity of his derivation of Stevin's law, he did not identify causes of hydrostatic phenomena other than weight. Stevin did not provide mechanistic explanations of hydrostatics whereas Pascal's concern to identify causes took him towards such explanations.

Pascal's style of argument is evident from the opening paragraph of his *Equilibrium of Liquids* where he describes some hydrostatic phenomena identified experimentally and offers the beginning of an explanation of them. The remainder of the Treatise is, in effect, an elaboration of his opening theme. It is worth quoting in full Pascal's opening paragraph, which would not be totally out of place in an introductory text today. The figures referred to are those in the first row of Pascal's Plate 1 which is reproduced in my Fig. 6.1.

> If one fastens to a wall several receptacles, one as in the first figure, the next sloping as in the second figure, another very wide as in the third, still another narrow as in the fourth, and the last merely a fine tube which ends in a broader but very short base as in the fifth, and if one fills them all with water to the same level, make apertures of the same area at the bases of each, and puts stoppers in to prevent the water from leaking out, experiment shows that it takes the same force to keep these stoppers in, although there are very different amounts of water in the various receptacles. This is because the water stands at the same level in all of them, and the measure of that force is the weight of the water contained in the first tube, which is uniformly of the same diameter throughout. If that water weighs a hundred pounds it will take a force of one hundred pounds to hold up each of the stoppers, even that in the fifth tube; though the water in it may weigh no more than one ounce. (Pascal 1937, p. 3)[3]

The phenomena described are to some extent counterintuitive, a fact emphasized by Pascal in his final clause where he notes that it is possible for an ounce of water to support a weight of a 100 pounds. But counterintuitive or not, *experiment shows* that it is the case. The ensuing treatise elaborates on and further applies the consequences of the brief explanation that Pascal proffers for this state of affairs.

Pascal did not start from definitions and postulates as Stevin had. Nowhere are there theorems or propositions derived mathematically. Plate 1 is the only figure in

[3] Most of my references to Pascal's Treatises are from the English translation in Pascal (1937). Where I have cause to refer to the original French I have used the edition of his works in Pascal (1970).

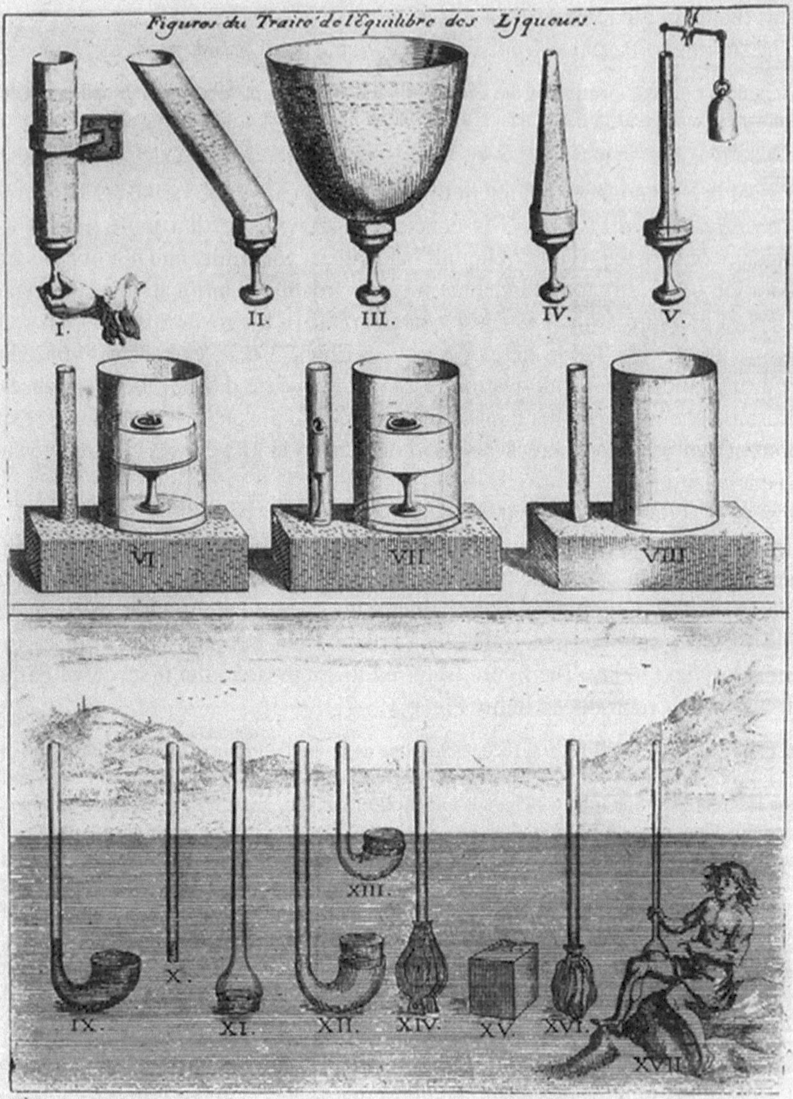

Fig. 6.1 Plate 1 from Pascal's *Equilibrium of Liquids* (Pascal 1937, p. 20)

The Equilibrium of Liquids. Geometrical figures of the kind that abound in Stevin's text are nowhere to be found in Pascal's. When Pascal did make a point with the aid of some geometrical reasoning he downplayed it, remarking that the proof that he had added to his deliberations 'will be understood by geometricians and may be disregarded by others'.[4]

In Pascal's hydrostatics, experiments played a key role in the articulation of the theory. He also construed them as providing evidence for it. Many of those experiments had been anticipated by Stevin. However, as we have seen, the latter confined descriptions of them to the *Practice of Hydrostatics* that was appended to the main treatise and which was deigned to present applications of, not evidence for, a theory established by other means. Pascal stressed the centrality of experiment in his mode of argument on the final page of the second of his two treatises, on *The Weight of the Air*. There he insisted that his Aristotelian critics 'should learn that experiment is the true master that one must follow in physics' (Pascal 1937, p. 75).

It is tempting for a modern reader to take Pascal's presentation of his hydrostatics as an *experimental* science at face value. The first paragraph of Pascal's treatise, reproduced above, encourages the idea that his hydrostatics was a physical theory borne out by experiment and which, as such, moved beyond Stevin's mistaken quest to present his hydrostatics as a mathematical theory on a par with geometry. Here I make two observations that will alert the reader to the fact that the situation was not quite so straightforward. First, a casual glance at the version of hydrostatics presented by Isaac Newton in Book 2 of the *Principia* is sufficient to indicate that its style more closely resembles that of Stevin rather than Pascal! Second, there are reasons to doubt that Pascal actually carried out the experiments he makes so much of in the *Equilibrium of Liquids*. We have work ahead of us to separate the substance from the rhetoric in Pascal's representation of his hydrostatics and to appreciate the significance of subsequent developments, especially those put in place by Robert Boyle and Newton.

In the remainder of this chapter I analyze the content of Pascal's hydrostatics and his case for it. I seek to explore the extent to which he fulfilled the aim, identified in the subtitle of his two treatises, to identify the causes of the behavior of liquids and air. I have already stressed the point that this task required the introduction of a cause or causes other than weight that would serve to capture the way in which liquids and air differ from solids. We know in retrospect that the addition of a technical sense of pressure as a cause to the action of weight was a crucial element of what was needed. The extent to which Pascal contributed to such an accomplishment raises matters sufficiently subtle that it will take me the remainder of this book to come to terms with them.

---

[4] Pascal (1937, p.8). I have altered the translation by Spiers and Spiers to render it more compatible with the original French, 'qui ne pourra être entendue que par les seuls géomètres, et peut être passée par les autres'.

6.3 Pascal's Theory of Hydrostatics

The most telling of the situations depicted in the first five figures of Pascal's Plate 1 is the fifth, my own version of which is shown in Fig. 1.1, p. 2, which indicates the challenging fact that it can take a 100 pounds or more to support an ounce of water. Pascal proceeded to elaborate on such a circumstance with a sixth situation. It involves a sealed container with two apertures in its upper surface, one large one small, into which vertical pipes are fitted. The pipe in the larger aperture is fitted with a piston, and Pascal raises the question of how much weight must be added to the piston to prevent it rising as a result of being pressed by water poured into the pipe in the smaller aperture. He stressed that the weight required to prevent the piston being raised by water in the narrower tube is dependent only on the height of the water in the latter and is independent of the diameter of the tube, so once again we have a circumstance in which it can take a weight of many pounds to support an ounce of water. As Pascal stressed, this sixth situation involves water pressing up on a stopper rather than weighing down on it. What is more, the same considerations would apply if the stoppered aperture were in the side of the vessel.

> The same thing happens when these stoppered apertures are at the side or even at the top of the broader base (Pascal 1937, p. 4).

Pascal concluded the first chapter of his treatise by formulating a rule governing the force necessary to prevent a stopper from being forced out in experiments of the kind that he has described.

> Whence it appears that the force necessary to prevent the water from escaping by an opening is proportional to the height of the water, and not to its expanse; and that the measure of the force is always the weight of all the water contained in a column as high as the water and of the same area as the opening.

This is in fact a rough statement of Stevin's law (rough because it does not take heed of the fact that the degree of pressing on a stopper that is not horizontal will vary with height). Pascal, like Stevin, has grasped the fact that water presses on a solid surface bounding it with a force that is independent of the orientation of that surface and is proportional to the height of the water and the cross-section of the surface. As far as Chap. 1 of Pascal's treatise is concerned, the version of Stevin's law with which it concludes is presented as a summary of what has been demonstrated experimentally to be the case. In the next chapter, a crucial one that introduces the substance of his theory, Pascal proceeds to explore what lies behind the experimental facts he has identified. In particular, and as the title of Chap. 2 makes clear, Pascal will seek to *explain* 'why liquids weigh in proportion to their height'.

Pascal articulated his theory by reference to the situation depicted in Fig. 6.2, my version of the seventh of the figures appearing in Pascal's Plate 1. The modern reader will recognize this as an idealized representation of the hydraulic press. However, there is nothing in the text to suggest that Pascal considered himself to be introducing a practically useful device here. It was one and a half centuries before the British inventor, Joseph Bramah, patented the first working version of a hydraulic

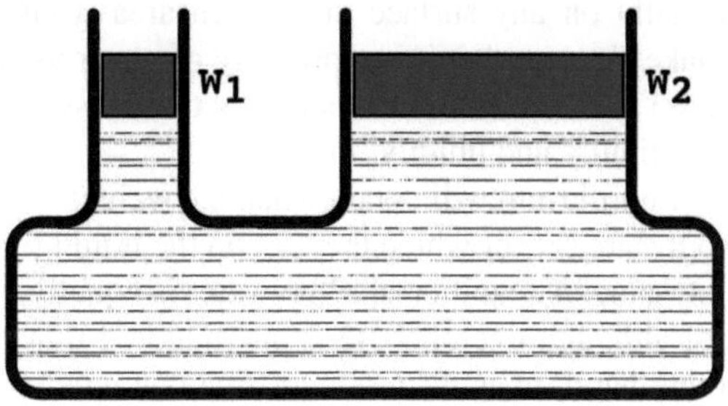

Fig. 6.2 The hydraulic press, anticipated by Pascal

press in 1795. The situation depicted in Fig. 6.3 served a theoretical purpose for Pascal, enabling him to develop the essentials of his theory of hydrostatics by reference to it. For convenience I will continue to refer to the device as the hydraulic press, a term not used by Pascal himself.

The effect of a weight on the water beneath it is described by Pascal as follows:

> Thus if a vessel filled with water has but one aperture, say one inch in area, and a piston is placed on it under a one-pound weight, that weight is exerted on every part of the vessel because of the continuity and fluidity of the water. To ascertain how much of the weight is borne by each part, the following rule holds good.
>
> Each area equal to that of the aperture, that is, one square inch, is pressed upon as if by a one-pound weight. (The weight of the water is not taken into consideration here since only the weight of the piston is being dealt with.) The one-pound weight presses the piston at the aperture, and each part of the vessel is more or less pressed in proportion to its area, whether that part be opposite the aperture, or to one side of it, or far, or near; for the continuity and fluidity of the water make all these circumstances equal and indifferent (Pascal 1937, pp. 7–8.).[5]

This crucial passage illustrates three key features of Pascal's hydrostatics. Firstly, we have a quantitative statement of the way in which the force exerted by W_1 appears as a force on other parts of the surface of the container. It is exerted as a force per unit area, one pound per square inch in this case. Secondly, we have a clear indication of the isotropic character of the force so transmitted, since the water presses

[5] I have altered the English translation to avoid using the noun 'pressure', thereby lessening the danger of mistakenly attributing to Pascal what is entailed by the modern, technical, usage of that expression. The French version of this crucial passage reads 'de sorte que si un vaisseau plein d'eau n'a qu'une seule ouverture, large d'un pouce, par exemple, où l'on mette un piston chargé d'un poids d'une livre, ce poids fait effort contre toutes les parties du vaisseau généralement, à cause de la continuité et de la fluidité de l'eau; mais, pour déterminer combine chaque partie souffre, en voici la règle. Chaque partie large d'un pouce, comme l'ouverture, souffre autant que si elle était poussée par le poids d'une livre (sans compter le poids de l'eau dont je ne parle pas ici, car je ne parle que du poids du piston) parce que le poids d'une livre presse le piston qui est. à l'ouverture, et chaque portion du vaisseau plus ou moins grande souffre précisément plus ou moins à proportion de sa grandeur, soit que cette portion soit vis-à-vis de l'ouverture ou à côté, loin ou près; car la continuité et la fluidité de l'eau rend toutes ces choses-là égales et indifférentes.' (Pascal 1970, p. 1046.)

equally on any surface of a given area whatever its orientation. Thirdly, Pascal makes it clear that it is a property of water as such (which it shares with other liquids) that lies behind the isotropic transmission of forces. It arises as a result of the 'continuity and fluidity' of water.

Equilibrium will obtain when the weights in the hydraulic press are proportional to the areas over which they act on the liquid for only then will the force on each be the same. At equilibrium 'the water is equally pressed upon under the two pistons; for although one of these is one hundred times as heavy as the other, it is, on the other hand, in contact with an area one hundred times greater' (Pascal 1937, p. 8). The quantitative measure of the pressing of the water, which is the same wherever it encounters a solid surface, is as a force per unit area.

A significant feature of Pascal's treatment of the hydraulic press is the extent to which it moves beyond effects arising from the weight of water. In the passage I have quoted, Pascal makes explicit the fact that he is concerned with the details of how the effects of the weights acting on the liquid are transmitted through it. 'The weight of the water is not taken into consideration here since only the weight of the piston is being dealt with.' Nor need the force on the liquid arise from weights. It could just as well arise from men pressing on the piston, a circumstance described by Pascal in the passage quoted below.

These reflections led Pascal to portray hydrostatic effects as governed by two 'principles' which he transported from the science of weight. The first of them involved the inverse proportionality of forces and distance moved when cause and effect are linked by a mechanical machine.

> It is remarkable that this new machine exhibits the same constant relation that is characteristic of all the old machines, such as the lever, the wheel and axle, the endless screw, and others, which is that the distance traversed increases in the same [*sic*] proportion as the force. For it is obvious that since one of these apertures [in the hydraulic press as illustrated in Fig. 6.2] is one hundred times as large as the other, a man pressing the small piston down a distance of one inch would move the other piston up only one-hundredth of that distance. It is the continuity of the water between the pistons that makes it impossible to move one without moving the other. It is clear that if the small piston moves one inch, the water thus moved presses on the other piston, and since the aperture of this one is one hundred times larger, it rises to only one-hundredth of the height. Thus the distances travelled are in the same [*sic*] ratio as the forces (Pascal 1937, p. 7).[6]

Pascal proceeded to draw implications from this that can be seen as an anticipation of the conservation of energy, a move that had been anticipated by Galileo.[7]

> Such may even be taken as the true cause of this effect, since it is evident that it amounts to the same thing whether we make one hundred pounds of water move through one inch or make one pound of water move through one hundred inches. Thus when one pound of water

[6] The context indicates that Pascal should have asserted the forces to be inversely rather than directly proportional to distance.

[7] In *On Mechanics,* when discussing the balance and the steelyard, Galileo noted that lifting a heavy weight by one weighing only a fifth as much situated five times further from the fulcrum of a steelyard was equivalent to lifting five times a weight five times smaller by means of a steelyard with equal arms. See Drabkin and Drake (1960, p. 159).

and one hundred pounds of water are so arranged that the one hundred pounds cannot move one inch without moving the one pound a distance of one hundred inches, they must stand in equilibrium, since one pound has as much force[8] to make the hundred pounds move through one inch, as the hundred pounds have to make one pound move through one hundred inches.

A second principle invoked by Pascal was one that had been made explicit by Torricelli (1644, Vol. 2, p. 105). It asserts that a mechanical arrangement will move spontaneously under the action of gravity only if such a movement results in a lowering of the center of gravity of the system as a whole. Pascal gives a geometrical argument to the effect that a small vertical displacement of the weights W_1 and W_2 in Fig. 6.2 about the equilibrium position will result in no change in the height of the system's center of gravity.

So Pascal's theory asserts that any force applied to a body of liquid will be transmitted through it in such a way that the force per unit area applied at one location will appear as the same force per unit area on any bounding surface of the body of liquid, whatever its orientation and however remote. Pascal stresses that such behavior is a consequence of the fluidity of the liquid by noting that, in the situations depicted in each of my Figs. 1.1 and 6.2, the support of a large weight by an ounce of water in the former and the lifting of a heavy weight by a light one made possible by the arrangement depicted in the latter, will not eventuate if the body of water is frozen. It is by virtue of its liquidity that a body of water is 'a mechanical machine for multiplying force' (Pascal 1937, p. 11).

Given the theory as summarized above, Pascal is in a position to show why liquids press to a degree that is proportional to the depth beneath their surface. The weights W_1 and W_2 depicted in 6.2 can be replaced by columns of water of the same weight. Since equilibrium requires that the weights be proportional to the areas of the surfaces on which they press it is clear that the water replacing the weights will be in equilibrium only if the columns of water have the same height. The situation depicted in Fig. 1.1 can also be interpreted by analogy with the hydraulic press in which one aperture is in the upper face and the other in the lower face of the body of liquid, as Pascal showed. The weight, W, presses up on the lower surface of the water, EF, and the column of liquid AB presses down on the much smaller surface at the base of the narrow tube, at B. The fact that equilibrium requires that the two pressings must involve the same force per unit area explains why the small weight of the liquid AB can support the much heavier weight, W.

> Again, the water in the small tube is in equilibrium with a one-hundred-pound weight. The reason is that the vessel at the base, which is wide but shallow, is a hermetically sealed vessel filled with water, with two apertures, one wide, at the bottom where the piston is, and the other narrow, at the top, where the small tube is. The water in the latter is virtually in itself a piston and balances the other because of the proportionality of the weights to the areas of the apertures; for as we have said, it does not matter whether the apertures face each other or not (Pascal 1937, p. 10).

---

[8] I here write 'force' for the original French '*force*', replacing the term 'efficacy' in the translation by Spiers and Spiers, the latter term being too readily interpretable in a way that suggests Pascal came closer to appreciating something like the conservation of energy than is warranted.

Any force applied to the surface of a body of liquid is transmitted isotropically through it because of its fluidity and appears as the same force per unit area on any other surface bounding the liquid. The weight of the liquid itself is just one of many possible ways in which a force can arise. This feature of Pascal's hydrostatics marks a major move beyond Archimedes and Stevin, both of whom confined their considerations to the hydrostatic effects of the *weight* of liquid alone.

6.4 Elaborations and Applications of Pascal's Hydrostatics

In the remaining five chapters of his short treatise Pascal detailed how his theory was borne out by a range of phenomena, some of them well known such as floating, others involving experimental facts known from the time of Stevin, and, to a limited extent, yet others that invoked novel experimental effects.

Pascal's theory can straightforwardly explain why water rises to the height of its source and why the height of water in tubes communicating via their lower ends is the same irrespective of how the tubes may differ in diameter (Chap. 3). A copper disc can be made to 'float' in water which is less dense than it. This can be achieved if the disc fits snuggly into the base of a narrow tube immersed vertically in water with its upper end open to the air. Provided the disc is sufficiently deep it will be supported by the water pressing against it from below (Chap. 4).

The isotropy of hydrostatic forces was reinforced in a number of ways. For instance, when a column of mercury is supported in a vertical tube immersed in water with its upper end open to the air, the height of mercury supported, which will be one fourteenth of the height of the water outside of the tube, is independent of whether the lower end of the tube points straight down, is bent at right angles or curves upwards so long as its open end is at the same depth in each case (Pascal 1937, pp. 12–13). A moderately inflated balloon will uniformly swell or contract as it is raised or lowered in water. 'This is because the weight of the mass of water above the balloon presses it on all sides towards the center' (Pascal 1937, p. 21). Pascal extended his discussion from inanimate objects to the effect of water on animals and humans deeply immersed in water. They are not distorted because they are pressed equally on every side (Chap. 7).

Armed with a theory that could yield the force exerted by a liquid on any solid surface confining it Pascal could explain floating and derive Archimedes' principle (Chap. 5).

> We have seen that water presses upward bodies that it bears upon from below, that it presses downward on those that it bears up from above, and that it presses to one side those that it bears upon from the opposite side. From this it can be readily inferred that, when a body is wholly submerged, then, since the water bears upon it above, below, and on every side, it strives to push it up, down and to all sides; but as its head is the measure of its force in all these efforts, there is no difficulty in determining which of them should overbear the rest. It is obvious at once that since the water has the same height on all the lateral faces, it will press upon them equally;—But as the water has a greater head against the bottom than against the top, it will obviously press the body more upward than downward; and since the

> difference between these heights of water is the height of the body itself, it will be readily understood that the water presses it upward and not downward, and with a force equal to the weight of a volume of water equal to that of the body (Pascal 1937, p. 16).

Here Pascal derives Archimedes' principle by considerations of the way in which weight acting downwards on a body is countered by the results of water pressing against it. By adding the pressing of the water to weight Pascal has, in this context, fulfilled the promise made in the subtitle of his treatises and identified the causes of the phenomenon in question. In my terminology, he here supplies a mechanistic explanation of the phenomenon of floating.

6.5 Pascal's Hydrostatics and Experiment

In this section I reflect on the relation between hydrostatics and experiment as it was reflected in Pascal's practice. There is much that is valid in the idea, sketched in Sect. 6.2, that Pascal presented a hydrostatics supported by experiment as opposed to one derived from unproblematic postulates according to the Euclidean ideal. However, the details of Pascal's case, as presented in *The Equilibrium of Liquids* do not altogether live up to his rhetoric concerning the importance of experiment.

There are grounds for doubting that Pascal actually performed many of the experiments he appealed to in the *Equilibrium of Liquids.* Boyle made a strong case to this effect in his commentary on Pascal's book, a work that we will be discussing in detail in Chap. 8. Boyle (1999, Vol. 5, pp. 206,) argued that stoppers in apparatus built according to some of Pascal's designs would leak, and that others were impracticable because they required glass tubes over 20 ft in length and vast quantities of mercury (pp. 224 and 255). Boyle also drew attention to the fact that some of Pascal's experiments required the observer to operate while 20 ft under water![9] It is safe to assume that Pascal did not actually perform a number of the experiments that he invoked with the implication that he had performed them. Pascal was also misleading insofar as he failed to mention that many of the experiments he did describe were in fact versions of those to be found in Stevin's *Practice of Hydrostatics.*

As far as the relation between theory and experiment in hydrostatics is concerned, there is much of significance that goes beyond the extent to which Pascal was guilty of misrepresentation. There is much that is valid and significant about Pascal's representation of his hydrostatics as an experimental science. In this section I attempt to articulate what was implicit in Pascal's hydrostatics in its historical context in that regard, drawing some distinctions which are significant but which Pascal himself did not draw.

As far as evidence supporting Pascal's hydrostatics is concerned, three different kinds can be distinguished. First there was common sense knowledge of liquids as distinct from solids. Second, there was experimental evidence generally accepted by

[9] Witness the gentleman in the last of the figures in Plate 1, reproduced in my Fig. 6.1.

the middle of the seventeenth century. Third, there was novel experimental evidence. All three kinds of evidence can be found at work in Pascal's *Equilibrium of Liquids.*

Pascal's appealed to his claims about the way in which a pressing at one location on the bounding surface of a liquid is transferred to other locations on the surface by virtue of the fluidity of the liquid in order to explain a range of common phenomena such as floating and the way in which water finds its own height. The phenomena explained were far from novel but the explanations offered were. Part of the case for Pascal's theory was the way in which it could naturally explain a range of common phenomena. As such, it was an empirical case.

As far as artificially produced experimental evidence is concerned, key experimental phenomena appealed to by Pascal had appeared in Stevin's *Practice of Hydrostatics*. For instance, the experiment illustrated in Fig. V of Pascal's Plate I originated with Stevin, with his discussion facilitated by a similar figure (Dijksterhuis 1955, p. 493). Also in Stevin's *Practice of Hydrostatics* is an anticipation of Pascal's experiment involving the 'floating' of a copper disc below the surface of water (Dijksterhuis 1955, p. 495). While it is not clear who actually performed these experiments, they had become accepted as experimental facts by the mid-seventeenth century. They were treated as such by Mersenne (1644, pp. 225–233) in his *Cogitata Physico-mathematica* and Isaac Beeckman and Descartes also treated some of the situations invoked by Stevin as experimental facts in need of explanation, as we saw in Sect. 5.3.

Claims for novelty could be made with respect to other experiments invoked by Pascal. His version of the hydraulic press and the way in which he exploited it theoretically was novel. Pierre Duhem (1905, pp. 604–605) has observed that Giovanni Battista Benedetti came close to appreciating the principle underlying the hydraulic press when reflecting on his practical experience with fountains, a printed version of his thoughts appearing in 1585. Duhem argued that Pascal may well have known of Benedetti's reflections either by reading him directly or through Mersenne. For his part, Mersenne (1644, pp. 228), in a work familiar to Pascal, proposed a thought experiment in which a stick is thrust into a hole in a movable lid covering the whole ocean. He considered how much force must be exerted on that lid to prevent the entry of the stick, concluding that it would exceed the force on the stick by a factor as large as that by which the cross section of the lid exceeded that of the stick; that is, that it would be enormous. Such anticipations notwithstanding, there is no doubt that Pascal's clear analysis of the principles underlying what we now call the hydraulic press and the theoretical use to which he put it was novel, whether he actually performed the relevant experiments or not.

Pascal's reflections on the changing size of a balloon thrust into water may have been novel as far as I know although it was a reflection of a kind that was common in the pneumatics that had become fashionable in the wake of Torricelli's experiment. I devote Chap. 7 to a discussion of such matters.

There is one experiment described by Pascal, involving a novel syphon effect which does seem to be an innovation of his. It is illustrated in Fig. 6.3. When the water surrounding the two mercury troughs rises sufficiently high for the mercury

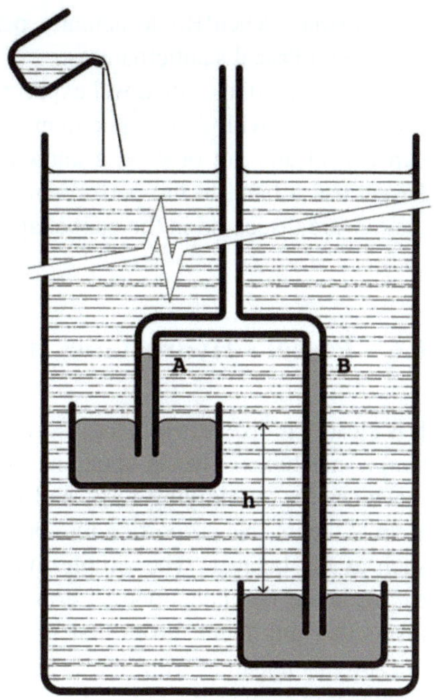

Fig. 6.3 A novel syphon effect. When the two columns of mercury meet in the parent tube the mercury in the upper trough empties into the lower one

columns in the tubes rising from them to meet in the parent tube, the mercury in the upper trough empties into the lower trough. The experiment is relevant to hydrostatics although a description of it appears in *On the Weight of the Mass of the Air*, where Pascal (1937, pp. 42–45) exploited it in his attack on the notion of the force of a vacuum. Ingenious though it may be, it is doubtful whether Pascal ever performed the experiment, which required a glass tube over 20 ft in length and large quantities of mercury. There is no record of the substantial coordinated effort that would have been involved in putting this experiment into effect.[10]

Novel experimental facts played a minor role at best in Pascal's hydrostatics. He could make a powerful case for his hydrostatics without them by showing how it could explain a wide range of common knowledge and commonly accepted experimental results. In this way he could demonstrate how liquids can function as a means for multiplying force and so could be confident that his hydraulic press would perform as required without ever constructing one. Another reason for downplaying the overall significance of the fact that Pascal did not perform all of the experiments he invoked is that, whilst not performed, versions of them were

[10]Alexander Koyré (1968, pp. 147–154) raised reasons to doubt whether Pascal actually performed experiments requiring such long tubes. Boyle redesigned Pascal' syphon experiment by replacing mercury with water and water with oil of turpentine. What is more, he actually performed the experiment and found it to live up to Pascal's expectations. Boyle's version of the experiment is described in Boyle (1999, Vol. 5, pp. 255–258).

performable. When Boyle actually performed them their outcome was in accordance with Pascal's anticipations.

As far as a reliance on novel experiments is concerned, the situation in hydrostatics contrasted with that in pneumatics. In the latter case, phenomena revealed by experiment and not otherwise known were of crucial importance. What is more, Pascal had a hand in the performance of some of the most important ones. Such matters are discussed in the next chapter.

6.6 Pascal's Hydrostatics in Perspective

It is clear that, like Stevin, Pascal saw hydrostatics as an extension of the science of simple machines. This is apparent from his talk of systems involving liquids as machines for multiplying force and from his recognition that such machines conform to the inverse proportionality principle exhibited by common simple machines such as the balance. The simple machines were intelligible insofar as, in them, causes could be understood as bringing about their effects by way of the mechanisms that link them. From that latter point of view, hydrostatic effects could be puzzling as apparent from the extent to which they were described as paradoxes. How is it possible for a body of water to exert a force in excess of its weight? How can water press horizontally against a containing wall and upwards against a floating body given that its weight acts vertically downwards? How is it that liquids can press around corners in a way that solids cannot? I wish to press the question of the extent to which Pascal's hydrostatics provided an answer to such puzzles.

Pascal did focus on a property of liquids other than weight, namely their fluidity. It was by virtue of that property of liquids, in addition to their continuity, that a force per unit area applied at one location on the surface of a body of liquid would be transmitted through the liquid and result in the same force per unit area on any other surface of the body of liquid, however remote and whatever its orientation. The pressing of a liquid characterized in this way supplied the cause that could be added to weight to make mechanistic explanations possible in hydrostatics. Pascal talked freely of liquids in equilibrium 'pressing' on all solid surfaces with which they are in contact with the same force per unit area. Pascal did not use the noun 'pressure' (*la pression*) in his treatise but he might well have. By appealing to the characteristic way in which liquids press on solid surfaces, Pascal could explain how hydrostatic effects come about to an extent that Stevin did not.

Pascal's theory stopped short of giving a complete mechanistic account of how forces are transmitted through liquids. He focused on the forces exerted on solid surfaces and did not explicitly address the issue of how forces act in the body of liquids. There were, however, some important moves in that direction. When he imagined that the weights acting on the two pistons in a hydraulic press be replaced by columns of water (thereby showing why the two water levels will be the same) then he had constructed a situation in which, instead of solid weights acting on the water there is *water* acting on water. The same can be said of Pascal's understanding of the situation depicted in Fig. 1.1 as an extension of the hydraulic press. Here the force of the weight acting up on the water in EF is countered by the force acting down on it by the *water* in AB. Again we have water acting on water.

There is another instance of Pascal coming close to Boyle's recognition of the way in which pressure is transmitted through liquids by way of liquid pressing against liquid. In the second of his Treatises, *On the Weight of the Mass of the Air*, Pascal (1937 pp. 70–71) contested the claim, still common in the first half of seventeenth century, that water does not weigh in water, claiming that he had already shown the contrary in *The Equilibrium of Liquids*. As a matter of fact, that matter is not discussed directly in the earlier work. However, it is quite easy to see how Pascal's discussion of the loss of weight of bodies immersed in water can be extended to suit his purpose. A body of water immersed in water will be pressed on all sides just like a solid body, and, what is more, the upward force on the lower surface will exceed the downward force on the upper surface because of the difference in height. Just as the apparent weight of the solid body is reduced by the weight of water equal to it in volume, so the apparent weight of a body of water is reduced by the weight of water that equals it in volume. That is, the apparent weight of water in water is zero because of the way in which its actual weight is countered by the upwards thrust. The pressing and the weighing combine to yield a mechanistic account of the situation, and implies the way in which forces are transmitted through liquids by way of the pushing of a portion of liquid against neighboring portions. What I have said here is a natural extension of Pascal's discussion, but is not one that he in fact took explicit advantage of.

If we appraise Pascal's hydrostatics from the point of view of the transition from the common to the technical sense of pressure then we can see that his moves beyond the common did not take him as far as the technical. This can be said with the benefit of hindsight. However, by pointing it out I set the scene for an appreciation of the way in which Boyle and then Newton were able to build on Pascal's efforts. Before turning my attention to those developments I digress somewhat from my task of tracing the evolution of hydrostatics to describe, in the next chapter, some important developments in pneumatics that took place around the same time and which have an important bearing on our story.

References

Boyle, R. 1999. *The works of Robert Boyle. 14 volumes*, eds. M. Hunter and E. Davis. London: Pickering and Chatto.

Dijksterhuis, E.J. 1955. *The principal works of Simon Stevin, Volume 1, Mechanics*. Amsterdam: Swets and Zeitlinger.

Drabkin, I.E., and S. Drake. 1960. *On motion and on mechanics*. Madison: University of Wisconsin Press.

Duhem, P. 1905. Le Principe de Pascal. *Revue Générale des Sciences Pures et Applique.* 16: 599–610.

Koyré, A. 1968. *Metaphysics and measurement: Essays in the scientific revolution*. London: Chapman and Hall.

Mersenne, M. 1644. *Cogitata Physico-mathematica*. Paris: Antoni Bertier.

Pascal, B. 1937. *The physical treatises of Pascal: The equilibrium of liquids and the weight of the mass of the air.* Trans. A.G.H. Spiers and I.H.B. Spiers. New York: Columbia University Press.

———. 1970. *Blaise Pascal: Ouevres Complète, Volume 2*, ed. J. Mesnard. Paris: Desclèe de Brouwer.

Torricelli, E. 1644. *Opera geometrica*. Florence: Amatoris Massae and Lauenti de Landis.

Chapter 7
Experimenting with Air

Abstract Developments in hydrostatics in the second half of the seventeenth century were influenced by developments in pneumatics. Those developments were very much informed by novel experiments carried out in contrived situations, beginning with Torricelli's famous experiment with what we now know as the mercury in glass barometer. That experiment was soon followed by others that exploited the space above the mercury in that device. It was shown that a partially inflated bladder expands when introduced into the space above the mercury and that a volume of air introduced into that space causes a much greater depression of the mercury than an equal volume of water. These latter experiments strongly suggested that air has a spontaneous capacity to expand that is a property distinct from weight. Once it is realized that air is distinguishable from solids by virtue of its capacity to expand then the question of how liquids differ from solids becomes an issue, since liquids, while sharing with solids and air the possession of weight, do not have a capacity to expand.

In this chapter I make a necessary digression to discuss developments in pneumatics that were triggered by Evangelista Torricelli's experiment in 1644 involving an inverted tube holding mercury and a number of related experiments that were soon to follow. These signaled the need to ascribe to air an inherent tendency to expand that was distinct from its weight. This circumstance helped to crystalize the question of how liquids differ from solids, since they lack the tendency to expand characteristic of air. I believe that it is against this background that we should read Robert Boyle's response to and deployment of Blaise Pascal's hydrostatics.

7.1 Torricelli's Experiment and Responses to It

The existence of air was a matter of common knowledge insofar as it can be felt and breathed. Forces arising from its motion were harnessed by sailing ships and windmills. The extent to which it can insinuate itself into any space, however small, and spontaneously expand to fill larger spaces made a distinction between it and both solids and liquids a matter of common sense. I have stressed in previous chapters

A.F. Chalmers, *One Hundred Years of Pressure*, Archimedes 51,
DOI 10.1007/978-3-319-56529-3_7

that a theorization of the distinction between solids and both liquids and air needed to move beyond weight, since liquids have weight and it was becoming increasingly apparent that air does too. It was by way of the experimental investigation of air, rather than of liquids such as water, that the first significant moves in that direction were made. From the 1640s onwards, and in a way that was facilitated by novel experiments, it became increasingly apparent that many effects brought about by the action of air were due to its capacity to resist compression in whatever direction and to spontaneously expand into any space available to it. In this chapter we will trace the emergence of the attribution to air of what Boyle was to refer to as its 'spring'.

The major impetus to the experimental investigation and theoretical characterization of air came from Torricelli's experiment, performed for him by Vincenzio Viviani in Florence in 1644.[1] In that experiment, the open end of a tube of mercury about three feet long was closed with a finger and immersed vertically, finger downwards, into a trough of mercury. The mercury in the tube fell until a column of mercury about 29 inches high remained supported in the tube, leaving a space at the top. Torricelli discussed his experiments in an interchange of letters with Michaelangelo Ricci in Rome, copies of which were sent by Ricci to Marin Mersenne in Paris. It was through Mersenne that the experiment became widely known and discussed, especially in France.[2]

The space above the mercury in Torricelli's experiment was the object of much attention and speculation. The fact that air had no way of entering the space during the inversion of the tube suggested that that space was indeed empty. That is, that it was a vacuum. This assumption was reinforced by Torricelli's observation, mentioned in his first letter to Ricci, that if water is allowed to rise above the mercury it completely fills the space, apparently uninhibited by the presence of anything else.[3] Advocates of the new mechanical world view seized on the apparent emptiness of the Torricellian space, interpreting it as a decisive refutation of the Aristotelian world view insofar as the latter involved a denial of the possibility of vacuum and the attribution of a range of hydrostatic and pneumatic effects, such as the operation of syphons and lift pumps, to nature's resistance to the formation of a vacuum.

The ways in which interpretations of Torricelli's experiment were deployed to argue for and against Aristotelian and mechanical world views were complex and indecisive. The mechanical philosophers were themselves divided on the question of the possibility of a vacuum. Those, like Pierre Gassendi, who favored an atomistic version of the mechanical philosophy in the manner of Democritus and Epicurus,

[1] Torricelli's experiment had been anticipated by Gasparo Berti in Rome about two years earlier. His version was with a long glass tube filled with water rather than mercury. Berti did not publish details of his experiment and repetition of it was to prove far from straightforward. See Webster (1965, pp. 445–446) for more detail and relevant references.

[2] The letters to and from Torricelli and Ricci are in Loria and Vassura (1919, Vol. 3, pp. 186–201). English translations are in Pascal (1937, pp. 163–170). All of my references are to the English versions.

[3] Pascal (1937, p. 165).

were fundamentally committed to a vacuum as that which separates and occupies the space between atoms. They could welcome the emptiness of the Torricellian space as a manifestation of the possibility of space devoid of matter. Rene Descartes, on the other hand, denied the possibility of empty space and considered the Torricellian space to be filled with the second element which, as discussed in Chap. 5, he considered to occupy the space between planets and suns and which was distinct from the third element of which the planets, including the earth, were composed. The Aristotelian world view could be related to the new pneumatic experiments in a variety of ways too. The mechanical philosophers tended to attribute to Aristotelians an undesirable anthropomorphic view that attributed various effects to nature's abhorrence of a vacuum, but it was by no means necessary that they adopt such a view. When Pascal argued against the 'force of the vacuum' theory he first cast it in a form that was not anthropomorphic and interpreted it as claiming that nature acted in such a way as to prevent the formation of a vacuum, a position similar to the one that Galileo adopted. Aristotelians could, and did, accommodate their world view to the new pneumatic experiments. For many of them, an essential property of air was its capacity to be rarefied, a capacity that it retained however much it had been in fact rarefied. From this point of view, the Torricellian space could be considered to be full of rarefied air, a view that could be reinforced by appeal to empirical evidence. For example, air bubbles were often seen to rise through the mercury and enter the space. When Boyle enclosed Torricelli's barometer in a chamber that he could exhaust with his air pump the mercury level in the barometer fell with each stroke of the pump until the level in the tube was only slightly higher than that in the trough. Boyle never could exhaust the air around the barometer to a degree that made the levels equal. It remained full of rarefied air!

Torricelli's experiment, and related ones that are identified below, were certainly deployed to support mechanical and Aristotelian world views and, more specifically, to defend specific stands on the possibility of a vacuum. But they did not need to be interpreted in that way. They can be interpreted as contributions to an attempt to construct *mechanistic* explanations of pneumatic effects, where 'mechanistic' retains its association with mechanical machines, rather than with the mechanical philosophy. Once the mercury in Torricelli's barometer is understood to be in equilibrium because the force of the atmosphere on the mercury in the trough is balanced by the weight of the column of mercury in the tube plus any force that might eventuate from whatever might fill the space at the top of the tube, then it makes little difference whether that latter contribution is zero or very small. The extension of mechanistic explanations of simple machines to pneumatics could proceed relatively independently of philosophical debates about world views and the possibility of completely empty space. As a matter of historical fact, the emphases on the philosophical dimension, that were prominent in the years immediately following Torricelli's experiment, gradually waned without being decisively resolved. What was eventually resolved was the way in which mechanistic explanations of the behavior of air could be given by adding a capacity to expand to its weight. In Chap. 8 we will see that when Boyle made crucial contributions to that quest at the experimental and theoretical level he explicitly formulated a version of the distinction I have made between the defense of world views and the search for mechanistic explanations.

In the following portrayal of pneumatics in the decade following Torricelli's experiment I focus mainly on the moves towards a theorization of air that made mechanistic explanations of pneumatic phenomena possible. If I bracket off the ways in which the various participants engaged in the broader philosophical or metaphysical debate, there is plenty that remains for me to investigate. Torricelli, Gilles Personne de Roberval, Pascal, and Jean Pecquet, among others, were centrally concerned with the project of articulating and defending mechanistic explanations of the behavior of air as revealed by their novel experiments.

7.2 Torricelli on the Weight and Rarefaction of Air

Torricelli's theoretical reflections on his experiment, revealed in his correspondence with Ricci, are as important as the performance of the experiment itself. In the first letter of June 11, 1644, in which Torricelli introduces Ricci to his experiment, the cause of the rise of the mercury in the tube is located in the weight of air for we 'live submerged in an ocean of the element air, which by unquestioned experiments is known to have weight' (Pascal 1937, p 164). There is a temptation to locate the cause of the suspension of the mercury in the space above it, the force resisting the formation of a vacuum. Torricelli argues against this on the grounds that if the space at the top of the tube is increased by adding a bulb that might be expected to increase the effect of the alleged cause, the mercury level remains unaltered. Torricelli maintains that the cause is 'external' and 'comes from without'.

> On the surface of the liquid which is in the basin, there gravitates a mass of air fifty miles high; is it therefore to be wondered at if in the glass [tube], where the mercury is not attracted nor indeed repelled, since there is nothing there, it enters and rises to such an extent as to come to equilibrium, with the weight of this outside air which presses upon it (Pascal 1937, p. 165).

The following week Torricelli moved beyond considerations confined to balancing weights when dealing with Ricci's query. In the first of those queries, Ricci imagined that the mercury in the trough be shielded from the weight of the air by means of a metal cap, suggesting that the fact that the mercury level in the tube remains unaffected implies that the weight of the air cannot be the cause of its rising. The first part of Torricelli's response is of no great consequence. If the metal cap is adjacent to the mercury in the trough then the failure of the mercury in the tube to subside is simply due to the fact that the mercury in the trough is unable to penetrate the solid metal. The second part of Torricelli's response, which assumes that the metal cap is placed so as to leave a layer of air above the mercury in the trough, is more consequential and focuses on the degree of compression of the air. If the air trapped between the mercury and the metal sheet retains the same density as it had before the sheet was set in place, then the mercury level in the tube will remain unchanged. Torricelli illustrated his position with an analogy with compressed wool. If a cylinder of wool rests on a solid surface and is compressed by a weight resting on its upper surface, that weight will be communicated to the solid surface via the compression of the wool. If a solid sheet of iron is now forced to cut its way through the wool so as to form a plane insulating the top half of the wool from the bottom half, the wool will continue to press on the solid surface to the same degree as before. The wool presses by way of its compression, which happens to have come about through application of a weight. Torricelli applied this line of argument to the weighing of air on the trough of mercury in the barometer experiment insisting that it is the weight of atmospheric air that causes the layer adjacent to the earth's surface to be compacted while it is the pressing of the air, so compacted, that results in a downwards force on the mercury surface. There is a move beyond mere weight considerations.

Ricci's next objection forced Torricelli to reflect on what is in effect the isotropy of pneumatic and hydrostatic forces. Ricci pointed out that the resistance of a closed syringe to the drawing out of the plunger is independent of the orientation of the syringe, suggesting to him that the effect could not be attributed to the weight of air, which acts downwards. In the first part of his response Torricelli related a story about the confrontation of a philosopher with his servant. The servant was engaged in fitting a faucet in the side of a wine barrel. This met with the scorn of the philosopher who insisted that 'the wine would never come out because the nature of weights is to press downwards and not horizontally from the sides'. Torricelli continued:

> But the servant made him see with his own eyes that although by nature liquids gravitate downwards, they press and spout in every direction, even upwards, as long as they find places to reach, – that is, places that resist with less force than their own (Pascal 1937, p. 169).

There is the hint of a significant move beyond weight considerations here, insofar as Torricelli acknowledges that liquids press and spout in every direction due to a force of 'their own'.

Torricelli also invoked a second example. If a pitcher is forced mouth-downwards into water and the upper surface of the pitcher is pierced, water rises into the pitcher, notwithstanding its weight.

Neither of Torricelli's responses quite address Ricci's query, which was related to the isotropy of the action of the air, since they involve the transmission of forces through liquids. The responses were relevant to an explanation of the barometer effect since the air pressing downwards on the mercury in the trough needs to be transformed into an upwards force supporting the weight of the mercury column. As the practical knowledge possessed by the servant in Torricelli's story implies, the fact that forces in liquids can press sideways was common knowledge. It was noted in the discussion of Stevin's hydrostatics that it had long been appreciated that water presses sideways on a containing wall. Torricelli noted and deployed that fact in his response to Ricci. In fact in the early 1640s he had conducted experiments to measure the rate of efflux of a liquid from low down in a filled container, horizontally and even vertically. However, as we saw in the previous chapter, it was left to Pascal to duly emphasize the isotropic pressing of liquids already implicit in Stevin's law.

Ricci's final query was also concerned with hydrostatics rather than pneumatics. It invoked the fact that a volume of liquid does not 'weigh' when immersed in a volume of like liquid: Water does not weigh in water. Torricelli's response was that such an observation is not relevant since the mercury in Torricelli's experiment is not immersed in anything. The upper surface of the mercury in the trough is open to the air, the vertical sides are constrained by glass surfaces and the upper surface in the tube is constrained by nothing. By pressing on the mercury in the trough the atmospheric air is able to support the column of mercury in the tube.

> We can only say that it is a fluid libratory body, one surface of which is bounded by a vacuum or near-vacuum which does not weigh anything at all, the other surface being bounded by air which is compressed by many miles of amassed air; and therefore that the surface which is not pressed at all rises because it is impelled by that other, and ascends until the weight of the metal raised comes to equalize the weight of the air pressing on the other side.

Torricelli has moved towards supplying a mechanistic explanation of the barometer effect. It is significant that he here attributes the pressing on the mercury in the trough to the compression of the air. However, the identification of the force that resists that pressing as due to a 'fluid libratory body', the mercury which is free to move up or down the tube until equilibrium obtains, goes little beyond common sense knowledge of the liquid state.

Our main focus in this chapter is on developments in the study of air rather than liquids. We follow that path further in the next section by considering some illuminating experiments that closely followed in the wake of Torricelli's.

7.3 Experiments with Air in the Wake of Torricelli

Freeing the question of the forces exerted by air arising from its compression from considerations of its weight was facilitated by experiments that built on Torricelli's and which exploited the space above the mercury displayed by it. One of them involved introducing equal volumes of water and air into the space, with dramatically different results.

The introduction of air caused a much greater depression in the mercury level than the introduction of water, notwithstanding the fact that the air weighed around a thousand times less than the water. Versions of the experiment were carried out independently, in 1647, by Etienne Noel, a catholic priest whose views were closely wedded to those of Descartes, and by Roberval, an active member of Mersenne's circle in Paris, who opposed Descartes and whose experimentation was not closely wedded to any particular philosophical world view. Roberval soon added a further experiment that illustrated in a striking way the hitherto inadequately appreciated capacity of air to expand. It involved the spontaneous expansion of a carp's bladder, freed of most of the air in it and tied at the neck, when inserted into the Torricellian space.

Roberval's experiments were known and widely discussed in the Mersenne circle and soon much more widely. His theoretical reflections as they appeared in his correspondence similarly became well known. In his explanation of the descent of mercury when air is admitted into the Torricellian space and of the expansion of a carp's bladder in that space, Roberval appeals to a 'power of rarefaction' possessed by the air. He moves towards a mechanistic explanation of the two effects by considering the equilibrium between that power of rarefaction plus the weight of the mercury column and the force of the atmosphere pressing down on the mercury in the trough. In the case of air admitted into the space, Roberval explains:

> But as a matter of fact, if besides mercury or water, there be admitted into any part of the tube some of our compressed and condensed air, as we have stated above, this air obtains its freedom and all its parts recoil and become rarefied and drive out the mercury or water, which for that reason will be depressed below the aforesaid height, either more or less, according to the air itself possesses greater or lesser power of rarefaction.[4]

The air that expands into the space presses in all directions, but it is only the mercury in the tube that is free to respond to the pressing, and so its level falls as a consequence. This is determined by the laws of mechanics and can be likened to the behavior of elastic solids, as in a bow.

> However, when I reasoned using the laws of mechanics, about the inducement of the depression of mercury by air it has not been possible to take up a satisfactory position or provide a better explanation according to the laws of nature, than if it were agreed that the air spontaneously and of itself became rarefied in the tube – yet it is not true to say that the whole of the air's force towards rarefaction is exhausted. The same air, while seeking to fill the whole space, exerts a pressure in all directions, on the adjacent bodies, the tube keeps in the particles from all other directions, mercury being the only one of them that can give way towards the lower part of the tube. Moreover, this explanation must be adapted: that the air which we respire only possesses such force towards dilation and rarefaction as is equal to the power of the natural element compressing or condensing it. Besides which, this also agrees with the laws of nature itself and applies in all other bodies which nature has granted powers of spring, as in bows, and innumerable other examples. All of which bodies, as long as they are compressed by force, but not extended beyond the limits of their own power, never cease to resist. They are carried by an innate force of resistance, which is the same as that force with which they are drawn or impelled by other bodies. So, at the beginning of its rebound the force is at its greatest, then it gradually grows less and less, and finally disappears altogether when the body has returned to its natural state (Webster 1965, p. 497–8).

The small amount of air in the carp's bladder expands in the Torricellian space because 'liberated at last from compression, being in a position where it is no longer compressed neither by our condensed air, not by other surrounding bodies, had expanded itself to the size which the bladder would permit'. (Webster 1965, p. 496) There is an equilibrium between the force by which the air in the bladder tends to

[4] The carp's bladder experiment and the one involving the insertion of air into the Torricellian space were described and analyzed in letters to 'Nobilem Virum Dominum des Noyers' in May–June, 1648. My quotations are from the English translation of those letters in Webster (1965, pp. 496–499). The passage cited is on p. 499.

expand and the force of contraction of the extended bladder skin plus force that might arise from anything that may be present on the Torricellian space. 'Doubtless, by the rules of mechanics, that air remaining in the bladder cannot be more rarefied than the rarefied air in the tube which surrounds the bladder.' (Webster 1965, p. 497).

What Roberval has made clear is that air, by virtue of being air, has a power to expand. That this power is quite distinct from its weight is reinforced by the comparison of that power with the forces arising from the distortion of elastic solids. Whether those latter forces were adequately theorized or not, it was a matter of common sense that they are distinct from weight. When a weight is suspended from a stretched string the weight is countered by the elastic restoring force. There are two differences between elastic forces as exhibited by air and those exhibited by solids. The first is acknowledged by Roberval. Compressed air tends to expand equally in all directions. A compressed or stretched solid contracts or expands mainly in the direction of the impressed force. There is a second difference that is implied but not stressed in what Roberval wrote. A sample of solid has a natural shape and size by virtue of being solid. It resists any force that tends to increase or decrease its shape or size. By contrast, a sample of gas has no natural shape or size. It naturally tends to expand into any space available to it.

Pascal was another major contributor to the novel experimental facts about the behavior of air. His account of new experiments on air published in 1647 included an account of Torricelli's experiment conducted with water rather than mercury.[5] The most famous of the experiments with which he was associated was that which demonstrated the fall in the level of a barometer when transported to the top of the Puy de Dôme. This was suggested by Pascal and carried out by his brother-in-law Perrier in 1648.[6] Two other experiments of significance were reported in *On the Weight of the Air.* One involved carrying a partially inflated balloon up a mountain, whereupon it became fully inflated. (Pascal 1937, p. 31) and the other was the 'void in the void' experiment in which the mercury level in a barometer situated within the void of a second barometer falls to zero (Pascal 1937, p. 56–57).

7.4 Pascal's *Treatise on the Weight of the Mass of the Air*

In *On the Weight of the Mass of the Air*, as in the *Equilibrium of Liquids,* Pascal's focus was on issues that could be explored and settled by appeal to experiment, and not on deeper metaphysical issues such as that concerning the possibility of a vacuum. In this he was following the likes of Stevin, Torricelli and Roberval rather than Descartes. He did enter into the metaphysical debate in his correspondence, siding with Gassendi and opposing the Cartesians on the possibility of a vacuum.[7] However,

[5] B. Pascal, '*Experience nouvelles touchant le vide*' in Pascal (1970, pp. 493–508).

[6] See Pascal (1937, pp. 97–112) for discussion of this experiment in the words of Perrier and Pascal.

[7] Pascal's views on the metaphysics of the vacuum are spelt out in his interchanges with the

in the treatises on liquids and on air Pascal was at pains to avoid metaphysics. In this section I first note how Pascal made this aspect of his endeavor explicit, before addressing the question of how his experimental program included a theorization of the nature of air.

Pascal considered himself to have decisively shown by appeal to experiment that his own account of the behavior of air was to be preferred to those that appealed to the force of the vacuum.

> I have recorded in the [Treatise on air] all the general effects that have been heretofore ascribed to nature's effort to avoid a vacuum, and have shown that it is utterly wrong to attribute them to that imaginary cause (Pascal 1937, p. 67).

He was not content to make his job easy by construing his opponents to be advocating nature's abhorrence of a vacuum.

> This manner of speaking is improper, since created nature, which is the nature under consideration, is not animated, and can have no passions. Such language is in fact metaphorical, and means nothing more than that nature makes the same efforts to avoid a vacuum as if she abhorred it. Those who use this phrase mean that it is the same thing to say that nature abhors a vacuum as to say that nature makes great efforts to prevent a vacuum.

Pascal made it clear that when he employed the term 'vacuum' it was to mean 'a space empty of all bodies which our senses can apprehend' (Pascal 1937, p. 68). The avoidance of the metaphysical dimension of debates about the vacuum implicit here was evident in Pascal's response to those of his opponents who claimed that only impure, not pure, air has weight.

> I am not acquainted with 'pure' air, and believe that it might be very difficult to find it. But throughout this treatise I am referring solely to air such as we breathe, regardless of its component elements. Whether it be compound or simple, that is the body which I call the air, and which I believe to have weight (Pascal 1937, p. 28).

As far as Pascal was concerned the purpose of his treatise on air was to settle by experiment the debate between his own view and that of those who attributed the behavior of air to the force of a vacuum. He considered the experiments that he was able to invoke to be incompatible with the force of a vacuum and to have shown that 'the weight of the mass of the air is their real and only cause'. (Pascal 1937, p.67) It is that latter clause that brings me to the main topic of this section. In his treatise on air Pascal was intent on explaining pneumatic effects solely by appeal to the weight of air. Absent from it is any attempt to capitalize on early attempts by the likes of Torricelli and Roberval to pinpoint a key property of air in addition to its weight, its innate capacity to expand. It would seem that the need in pneumatics for an elaborated version of Roberval's 'power of rarefaction' was far from obvious.

When Pascal wrote *On the Weight of the Mass of the Air* around 1654 he could draw on the experimental investigations we have described above, including his own, and the theoretical reflections that they engendered. As the title already

---

Cartesian Étienne Noel, reproduced in Pascal (1970, pp. 509–540). Also of particular relevance is Pascal's letter to Le Paileur in Pascal (1970, pp. 556–576). Koyré (1968, p. 155) has noted Pascal's debt to Gassendi as far as his position on the vacuum was concerned.

indicates, Pascal's emphasis was on the explanation of pneumatic effects by appeal to the weight of air. As we saw in the last chapter, in the *Equilibrium of Liquids* Pascal made considerable progress towards capturing the distinctive characteristic of liquids by highlighting their fluidity in a way that marked a significant move towards the modern concept of pressure. This was not matched by his theorization of air in his pneumatics. In the first chapter of *On the Weight of the Mass of the Air* Pascal made it clear that his intent was to explain a wide range of experimental effects by appeal to the weight of air rather than the force of the vacuum. He noted that atmospheric air presses on the earth by virtue of its weight, just as the oceans do and acknowledged that the degree of pressing is dependent on height, just as is the case with water. Pascal (1937, p. 32) continued the analogy with liquids by noting that 'bodies are pressed on all sides by the weight (*le poids*) of the air above them' drawing a comparison with what had been shown in the *Equilibrium of Liquids*. The reference to 'pressing on all sides' alludes to the isotropic character of the pressing, although the centrality of the notion is obscured by use of the term 'weight'.

7.5 Pecquet and Air's Elater

Jean Pecquet was a French physiologist and anatomist of some distinction who followed in the experimental tradition set by William Harvey. From 1647 he developed an interest in the new experiments being performed and discussed at that time and participated in the discussions and demonstrations taking place in the Mersenne circle in Paris. Pecquet included a chapter describing the new developments in pneumatics in a book on physiology, *Experimenta Nova Anatomica*, published in 1651. It was translated into English in 1653 and became one of the main routes by which the new trends in pneumatics were transmitted to the likes of Henry Power and Boyle in England.

It is no accident that an anatomist like Pesquet was involved in developments in mechanics and the use of experiment. He was one of those physiologists mentioned in Sect. 1.6 and discussed in Bertoloni Meli (2011) who saw themselves as mechanists insofar as they thought of an organism as a machine whose properties arise as a result of the properties of its parts and the way in which those parts are assembled. Just as the workings of a clock can be best revealed by taking it apart to identify its parts and the way in which they are assembled, so anatomy requires dissection and the use of microscopes to serve a similar end. An anatomist like Pesquet was both a mechanist and an experimental philosopher and, as such, had reason to attend to and collaborate with those in the Mersenne circle more centrally concerned with physical mechanics.

Pecquet did not perform pneumatic experiments of his own but gave clear accounts of the key ones that we have described in this chapter. In addition he built on Roberval's notion of the 'power of rarefaction' possessed by air, introducing his own term, 'elater', to describe it. He made it clear that elater was responsible for the behavior of air alongside and in addition to its weight.

> It is my suggestion to you that this air is like spongy or more woolly heaps lying on the matter of the Terraquaceous Globe, as a consequence of which each successively higher layer compresses the lower; they are sustained so that, the nearer the layers are to the earth, then also they are more closely compressed by the weight and pressure of those lying on them. On account of the spontaneous dilation, which I call Elater, however strongly they are compressed by this cumulative burden, if set free the air rarefies.
>
> Hence, I infer that such lower parts are subjected to the whole burden, so that of all parts they have the maximum degree of condensation. Because of this same cause, whereby it exerted its powerful tendency to rarefy, not only by means of its weight but also by its elater pressing against the surface of the Terraquaceous globe.[8]

The recognition that elater is a key property of air in addition to its weight is again stressed in Pecquet's heading for a section describing the markedly different effects arising from the introduction of water and air into the Torricellian space. 'Water only compresses the Terraquaceous globe by virtue of its weight, but air not only by its weight but also by Elater.'[9] Solids, liquids and air alike possess weight, whereas, of the three, only air possesses elater.

Two important points are evident from Pecquet's insightful summary of the state of play in pneumatics around the middle of the seventeenth century. Firstly, the phenomena dealt with were revealed by experiment and not otherwise appreciated. Secondly, an explanation of those phenomena needed to appeal to a property characteristic of air, namely, its capacity to expand into any space open to it. It became clear that that capacity was distinct from the air's weight. In the next chapter we will see how Boyle built on those features of pneumatics and drew lessons from them that spawned important advances in hydrostatics.

References

Bertoloni Meli, D. 2011. *Mechanism, experiment, disease: Marcello Malphighi and seventeenth-century anatomy*. Baltimore: The John Hopkins University Press.

Koyré, A. 1968. *Metaphysics and measurement: Essays in the scientific revolution*. London: Chapman and Hall.

Loria, G., and G. Vassura. 1919. *Opere di Evangelista Torricelli*. Vol. 3, 186–201. Florence: Montanari.

Pascal, B. 1937. *The physical treatises of Pascal: The equilibrium of liquids and the weight of the mass of the air*. Trans. A. G. H. Spiers and I. H. B. Spiers. New York: Columbia University Press.

———. 1970. *Blaise Pascal: Ouevres complète*, ed. J. Mesnard, vol. 2. Paris: Desclèe de Brouwer.

Pecquet, J. 1651. *Experimenta nova anatomica*. Paris: S. and G. Cramoisy.

Webster, C. 1965. The discovery of Boyle's law and the concept of the elasticity of air in the seventeenth century. *Archive for History of Exact Sciences*. 2: 441–499.

[8] As cited in Webster (1965, p. 454).

[9] Webster (1965, p. 499).

Chapter 8
Boyle on Mechanism and Pressure

Abstract Boyle made explicit the notion of mechanism that was implied in the science of simple machines and which he brought to bear on his pneumatics and hydrostatics. Machines bring about their effects by virtue of the way in which pushes and pulls are transmitted through them via the connection of their parts. Boyle also made explicit the point that an understanding of machines involved taking properties such as weight and elasticity for granted, without requiring that they be explained at some deeper, corpuscular or other, level. Boyle developed his notion of the spring of air to capture its inherent tendency to expand and exploited it to give mechanistic accounts of a range of pneumatic phenomena, some of which were produced with the aid of his newly devised air pump. When Boyle became familiar with Pascal's *Equilibrium of Liquids* soon after its publication in 1663 he turned his attention to hydrostatics. As well as illustrating its fundamentals by a range of easy to perform experiments he made an important theoretical innovation. He considered pressures within the body of liquids, acting on either side of imaginary planes of various orientations. This was a move that Newton was to take for granted and exploit to great effect.

8.1 Introduction

When Robert Boyle began his serious involvement in experimentation from the mid-1650s he was aware of most of the developments in pneumatics described in the previous chapter. He took them further by elaborating on Gilles Personne de Roberval's 'power of rarefaction' and Jean Pecquet's 'elater', fashioning his own concept of the 'spring' of air and vastly extending the experimental case for what could be explained by appeal to it. He did this by employing an air pump designed for that purpose. The results of his experimental and theoretical labors in this respect appeared in a lengthy tract published in 1660. It was not until Boyle read Blaise Pascal's two Treatises in 1664, a year after they were published, that Boyle gave serious attention to hydrostatics. Here too his contributions were theoretical as well as experimental. In the work on hydrostatics that Boyle published in 1666 one finds an important transformation of the common sense of pressure in the direction of the

Much of the material in this chapter has appeared in Chalmers (2015b).

A.F. Chalmers, *One Hundred Years of Pressure*, Archimedes 51,
DOI 10.1007/978-3-319-56529-3_8

modern, technical sense. Boyle was able to provide mechanistic explanations of some otherwise puzzling hydrostatic phenomena employing his new concept of pressure. Those are the developments I describe in this chapter.

8.2 Mechanical Machines and Mechanistic Explanations

By the middle of the seventeenth century the science of weight, which had important origins in the works of Aristotle and Archimedes, had undergone significant developments on a number of fronts. The theory of simple machines, such as the balance and the pulley, had reached a considerable degree of sophistication. The kinematics of motions caused by weight had been explored by Galileo and clarified by his successors. As far as dynamics is concerned, moves beyond the medieval impetus theory involved the notion of inertial motion, centrifugal forces involved in circular motion and attempts to formulate laws of impact. Because the focus of this book is on the emergence of hydrostatics I will restrict my considerations to what we would now call statics, the theory of equilibrium exemplified by simple machines such as the balance. As I have urged in a number of places earlier in this book, it is appropriate to view the science of hydrostatics as an extension of, and as growing out of, the theory of simple machines.

A review of the key features of simple machines will serve as useful background to Boyle's contributions to pneumatics and hydrostatics. The functioning of a simple machine depends on the properties of its parts and the distinctive way in which those parts are connected together to form a mechanism. Central to that functioning is the forces that drive the machine, weight in the case of the balance, elasticity in the case of a watch. The constraints imposed on machines by the properties of their parts and their mode of connection, the rigidity of gear wheels, the constant length of pulley strings and so on, are also crucial. If hydrostatic devices are to be understood as machines then the precise way in which fluids constrain the system of which they form a part and serve to transmit forces from one point to another needs to be specified.

Given my construal of the situation, Boyle's problem situation in pneumatics and hydrostatic can be construed as follows: How can the common sense distinction between solids, liquids and air be made precise in such a way that equilibrium in systems involving liquids and air can be understood in the same way that simple machines are? How can machines involving liquids and air be comprehended as mechanisms? Such a characterization of the situation is not simply an imposition of my own perspective. Many of the points I have raised were made by Boyle himself. In doing so he made explicit what had been implicit within the science of simple machines and attempts to refine and extend it that he had inherited from his predecessors. The key reason why Boyle needed to make them explicit was the need to distinguish mechanistic explanations from the ultimate explanations involved in the mechanical philosophy that had made its appearance by the mid-seventeenth century, a point introduced in Sect. 1.6.

8.3 Boyle on Mechanism and Intermediate Causes

In the year following the appearance of *Spring of the Air* Boyle published *Certain Physiological Essays and other Tracts.* In two of those essays, 'A proemial essay touching experimental essays in general' and 'Some specimens of an attempt to make chymical experiments useful to illustrate the notions of the corpuscular philosophy', Boyle addressed the question of the nature and purpose of the experimental enquiry that he had been engaged in for the previous half decade. He also introduced the mechanical or corpuscular philosophy that he was later to spell out, advocate and help to make fashionable. Here Boyle made a number of the points about mechanism the importance of which I am keen to stress.

In the latter of the essays mentioned above, Boyle claimed, as he was to repeatedly insist thereafter, that the mechanical philosophy, according to which the material world is a system of interconnected atoms or corpuscles, is intelligible in a way that the Aristotelian appeal to forms and qualities is not. He justified his use of the term 'mechanical philosophy' on the grounds that corpuscular mechanisms are analogous to those that 'are obvious and very powerful in mechanical engines' (Boyle 1999, Vol. 2, p. 87).[1] In each case mechanisms are comprehended in terms of the contact action between the parts of which they are composed. Here Boyle trades on the extent to which the working of simple machines is obvious in a common sense. One might well question whether the analogy between simple machines and corpuscular mechanisms is close enough to warrant the intelligibility of the former to be transferred to the latter. This is especially so since notions such as rigidity, weight and elasticity, vital for the understanding of simple machines, cannot be applied at the corpuscular level at Boyle's own insistence. But that quibble aside, I note, and certainly am prepared to grant, Boyle's insistence that the workings of simple machines are intelligible in a common sense. As Boyle says, they are 'obvious and very powerful'.

In the 'Proemial essay' Boyle elaborated on his position in a way that reinforced a distinction between simple mechanisms and corpuscular mechanisms. He insisted that the kinds of knowledge that can be vindicated by experiment involves what he sometimes referred to as 'intermediate causes' such as weight and elasticity. These are distinct from the ultimate atomic or corpuscular causes to which, as a strict mechanical philosopher, Boyle believed such properties as weight and elasticity were to be reduced.[2] Boyle insisted that there is merit in knowledge that explains by appealing to unexplained intermediate causes.

---

[1] Boyle referred to machines like the balance, his air pump and the clock on Strasbourg cathedral as 'engines'.

[2] Boyle referred to 'intermediate' causes in "A proemial essay' published in 1961. This was an introduction to *Certain Physiological Essays* in which Boyle reported on some of his early experimentation other than his pneumatic experiments. It is reproduced in Boyle (1999, Vol. II, pp. 9–34). The reference to intermediate causes is on p.23. Intermediate causes are characterized more fully in a manuscript fragment dealing with final causes and the uses of experiment dating from around 1688 and reproduced in Boyle (1990 Volume IX, f40–41, reel 5, frame 250). 'Of the subordinate

> And though it must not be denied that it is an advantage as well as a satisfaction to know in general how the qualities of things are deducible from the primitive affections of the smallest parts of matter, yet whether we know that or no, if we know the qualities of this or that body they compose, and how 'tis disposed to work on other bodies or be brought on by them, we may, without ascending to the top of the series of causes, perform things of great moment and such as without the diligent examination of particular bodies would, I fear, never have been found out a priori even by the most profound contemplators (Boyle 1999, Vol. II, p. 24).

Knowledge of intermediate causes is to be found by 'diligent examination of particular bodies', that is by experimenting, rather than by a priori contemplation. Such knowledge, which is not otherwise available, is of great moment and not to be despised since it is explanatory.

Boyle explicitly cites gravity as an intermediate cause that can legitimately be appealed to in experimental science.

> He gives some reason why stones and iron, and all other heavy bodies, will swim in quicksilver, except gold, which will sink in it; that teaches that all those other bodies are *in specie* (as they speak) or bulk for bulk, lighter than quicksilver; whereas gold is heavier. He, I say, may be allowed to have rendered a reason of a thing proposed, that thus refers the phenomenon to that known affection of almost all bodies here below, which we call gravity, though he does not deduce the phenomenon from atoms, nor give us a cause of gravity; as indeed scarce any philosopher has yet given us a satisfactory account of it (Boyle 1999, Vol. 2, p. 22).

Boyle's general stance on the merit of mechanistic explanations and on the legitimacy and usefulness of appeal to intermediate causes was explicitly put to use in his texts on pneumatics and hydrostatics. Boyle made one of his most forthright expressions of the sense in which he regarded his accounts there to be 'mechanical' in *An Hydrostatical Discourse*, composed in 1672 as a response to criticism from Henry More. Defending his appeals to the weight and spring of the air in his pneumatics, Boyle wrote:

> And since such kinds of explanations have been of late generally called mechanical, in respect of their being grounded upon the laws of the mechanics, I, that do not use to contend about names, suffer them quietly to be so: And to entitle my now examined explication to be mechanical, in the usual sense of that expression, I am not obliged to treat the cause of gravity in general; since many propositions of Archimedes, *Stevinus* and those others that have written of statics, are confessed to be mathematically or mechanically demonstrated, though those authors do not take upon them to assign the true cause of gravity but take it for granted, as a thing universally acknowledged, that there is such a quality in bodies that they treat of. And if in each of the scales of an ordinary and just balance, a pound weight, for instance, be put; he that shall say, that the scales hang still in equilibrium because the equal weights counterpoise one another; and in case an ounce be put into one of the scales, and not in the opposite; he that shall say that the loaded scale is depressed, because it is urged

---

or intermediate causes or theories of natural things; there are many; some more or less remote from the First Principles and yet each of them capable to afford a just delight and useful instruction to the mind. And these we may call the cosmographical, the hydrostatical, the anatomical, the magnetical, the chemical and other causes or reasons of phenomena as those which are more immediate (in our way of estimating things) than the general and primordial cause of natural effects.' See also Boyle (1990, Vol. VIII, f184, reel 5, frame 189).

> by a greater weight than the other, will be thought to have given a mechanical explanation of the equilibrium of the scales, and their losing it; though he cannot give a true cause, why either of those scales tends towards the center of the earth (Boyle 1999, Vol. VII, p. 148).

This passage shows all of the features of mechanism reviewed in Sect. 8.2. The lowering of one weight causes the other to rise because they are materially connected by a rigid arm, balance occurs when the tendency of a weight to move downwards is counteracted by its being urged upwards by the force of the counterweight. This understanding does not require that gravity itself be explained.

The standpoint adopted by Boyle and described in the foregoing is clearly implicit in *Spring of the Air*, where Boyle (1999 Vol. I, p. 166) insists that his business 'is not – to assign the adequate cause of the spring of the air but only to manifest, that the air has a spring, and to relate some of its effects'. The spring of the air was treated as an intermediate cause. When, in a later work, Boyle turned his attention to hydrostatics he insisted that that field qualified as 'philosophy'.

> Hydrostatics is a part of philosophy, which I confess I look upon as one of the ingeniousest [sic.] doctrines that belong to it. Theorems and problems of the art, being most of them pure and handsome productions of reason duly exercised on attentively considered subjects, and making in them such discoveries as are not only pleasing, but diverse of them surprising, and such as would make one at first wonder by what kind of ratiocination men came to attain the knowledge of such unobvious truths. – For there are many things, as well as the more familiar, as of the more abstruse phenomena of nature that will never be understood by those that are strangers to the hydrostatics (Boyle 1999, Vol. V, p. 194).

This passage appears early in a work that proceeds to describe phenomena explored by *experiment* and explained by appeal to the weight and fluidity of water. If philosophical knowledge is distinguished from that of artisans by the fact that the former, unlike the latter, identifies causes which facilitate explanations of phenomena, then Boyle can insist that hydrostatics is to be deemed part of philosophy. This is in spite of the fact that the causes and explanations involved are intermediate rather than ultimate ones.

Pneumatics involved a move beyond the familiarity with weight to novel phenomena revealed by experiment. As a consequence, novel notions other than weight needed to be fashioned and the adequacy of various claims made with their aid needed to be established. This needed to be done not by appeal to their unproblematic character but by appeal to their explanatory power and the extent to which their use was illustrated and supported by experiment. Hence Boyle's reference in the passage quoted above to 'surprising' discoveries, and 'unobvious truths' revealed by 'diligent examination of particular bodies' and requiring for their comprehension novel concepts that are 'handsome productions of reason'. His advance into such territory involved his fashioning of the concept of the spring of the air as a mechanical cause distinct from weight. At a later date he came to recognize that hydrostatics needed some other advance beyond weight, since liquids do not possess a significant spring. It was to meet that need that Boyle transformed the common sense of pressure into something approaching the modern, technical concept.

Unlike his predecessor Simon Stevin, Boyle was not particularly concerned to express his accounts of pneumatics and hydrostatics mathematically. Rather he aimed to identify the intermediate causes of the phenomena and to show how they functioned mechanically rather than mathematically. By doing so he aimed to provide mechanistic explanations of hydrostatics and pneumatics that could be illustrated and vindicated by a range of experiments. He aimed for 'physico-mechanical' rather than 'physico-mathematical' explanations. This is the point of view I bring to bear on my analysis of Boyle's treatment of spring, fluidity and pressure.

8.4 Boyle's Innovations in Pneumatics

Boyle's serious involvement with experiment began with his move to Oxford in 1655. There he conducted a range of experiments, mainly in chemistry and pneumatics.[3] News of the new pneumatic experiments from Torricelli onwards reached England by a number of routes and we know that Boyle had up-to-date knowledge of them at least through his reading of Pecquet's *Experimenta Nova Anatomica*.[4] Boyle was inspired to conduct pneumatic experiments of his own once he heard of the air pump devised by Otto von Guericke around 1654. In 1658 he arranged for his assistant, Robert Hooke, aided by Ralph Greatorex, a London instrument maker, to build him an air pump that would enable him to conduct experiments in a 'receiver' from which air had been pumped. Boyle's description of the pump, a range of experiments conducted with it and his deployment of the 'spring of the air' to explain their outcomes were described in *New Experiments Physico-mechanical Touching the Spring of the Air and its Effects'*, published in 1660 (Boyle 1999, Vol. 1, pp. 143–300).

While the deployment of the spring of the air by Boyle in his pneumatics was not entirely novel it was highly significant nevertheless. The fact that the concept needed to be elucidated, clarified and publicized is evident from the fact that, as we have noted, the treatise on air written by Pascal in 1654 gave undue emphasis to the weight of air and did not involve a concept playing the role of Boyle's 'spring'. When Boyle came to review the book embodying Pascal's treatises, on liquids and on air, following their publication as one volume in 1663, he paid scant attention to the latter, concentrating his review on *Equilibrium of Liquids*. The reason he gave was that Pascal's treatise on air had been rendered outdated because of the extent to which the matters had been 'more clearly made out by experiments, which could not be made by *Monsieur Pascal*, and those other learned men, that wanted the advantage of such engines and instruments, as have in this place [The Royal Society] been frequently made use of'.[5] But Boyle may well have added that it was not only the character of the experimental evidence he invoked that rendered Pascal's theory

[3] For details of Boyle's initiation into the experimental life see Michael Hunter (1995, 59–103), William Newman (2003, pp. 54–91) and Newman and Principe (2002).

[4] For an account of the introduction of Continental pneumatics into England see Webster (1965, pp. 454–464). Boyle cites Pecquet a number of times in his tract of 1660.

[5] Boyle (1999, Vol. V, p. 205).

outmoded, but the latter's failure to theorize the characteristic feature of air, over and above its weight, that was responsible for pneumatic effects and which Boyle had termed its spring. While Boyle's major contributions to pneumatics came via his experiments, especially but not exclusively those utilizing the air pump, he also deployed and reflected on them to clarify and extend its theoretical formulation.

In his explication of the spring of the air Boyle followed his predecessors by drawing analogies between air and elastic solids. Boyle's theorization of the spring of the air made explicit the distinction between elasticity in solids and in air. An elastic solid resists expansion as well as contraction whereas air has 'a restless power of expansion' only.[6] What is more, this power of expansion is isotropic, differing from elasticity in solids in that respect. These differences between air and solids were stressed by Boyle, for instance, when he described the force experienced on a finger used to seal the upper end of a barometer tube.

> It is to be considered, that the air being a fluid as well as a heavy body, it does not (as grosser weights would) press only on the upper parts of the finger, but pressing as much of the finger as is exposed to it almost everywhere, and almost uniformly as well as strongly, it does by its lateral pressure on every side thrust in the pulp of the finger into the hole where there is not any resistance at all, or at least near so much pressure against the pulp as that of the ambient air against the parts of the finger ambient to it. (Boyle 1999, Vol. III, pp. 22–23).

I have selected this quotation from the many other references implying the isotropy of the spring of the air because here Boyle makes it clear that it is the fluidity of air as opposed to the solidity of 'grosser weights', that is responsible for the isotropy.[7]

The passage quoted above is taken from Boyle's *Defense against Linus,* published in 1662 as a response to a critique of the *Spring of the Air* by the Aristotelian Franciscus Linus. The latter proposed that forces attributed by Boyle to the spring of the air were in fact due to the stretching of a 'funiculus' (thread) joining the extremities of evacuated regions. Hence, for instance, the force on the finger stopping the barometer tube described in the passage in question is indeed due to a sucking, in keeping with what might well be inferred, and indeed felt, when the experiment is conducted by the uninitiated. Boyle was keen to establish that pneumatic forces arise from the pushing due to condensed air rather than the sucking due to expanded air. Air does not have a natural propensity to contract in the way that a stretched solid does. Forces resisting expansion are plausible in solids, which are relatively rigid and have a natural size. However, the fluidity of air and its propensity to expand imply that its parts are not connected in a way that would generate a resistance to expansion. While clockwork mechanisms can be driven by the pushes and pulls arising from the contraction or stretching of a metal spring, Boyle was intent on showing that pneumatic effects arise as a result of pushes only, those due to the 'restless power of expansion' that he called its spring.

---

[6] Boyle (1999, Vol. I, p. 245).

[7] About a decade later Boyle devoted a whole tract to the differences between pressure in solids and in fluids. See Boyle (1999, VII, pp. 215–225).

Boyle strengthened the case for attributing the causes of pneumatic effects to pushes arising from the spring of air by correlating the strength of experimental effects with variations in the spring that he was able to bring about in the receiver of his air pump. He was able to give mechanistic explanations of a wide variety of effects, some of them previously established and some of them novel fruits of his own experimentation. So, for instance, the difficulty of lifting the stopper in the receiver of the air pump once a significant quantity of air is pumped from the latter is explained by the fact that

> the air in the receiver, being very much dilated, its spring must be very much weakened, and consequently it can but faintly press up the lower end of the stopple, whereas the spring of the external air being no way debilitated, he that a little lifts up the stopple must with his hand support a pressure equal to the disproportion between the force of the internal expanded air, and that of the atmosphere incumbent upon the upper part of the same key or stopple (Boyle 1999, Vol. I, p. 171).

Likewise, the expansion of a bladder caused by the evacuation of the receiver in which it is housed is due to the 'surmounting of the debilitated spring of the ambient air remaining in the vessel, by the stronger spring of the air remaining in the bladder' (Boyle 1999, Vol. 1, p. 174). Two polished discs of marble in contact, one beneath the other, provides a third from amongst the many examples that can be drawn from Boyle's text. The difficulty of separating the discs is due to the fact that the lower surface of the lowermost stone 'being freely exposed to the air is pressed upon by it, whereas the uppermost surface, being contiguous to the superior stone, is thereby defended from the pressure of the air which consequently, pressing the lower stone against the upper, hinders it from falling' (Boyle 1999, Vol. I, p. 238). The transmission of the force due to air pressing on a water surface through the body of the water is demonstrated by immersing a partially inflated bladder beneath the water surface. One then demonstrates how the balloon expands or contracts as the degree to which the air presses on the water surface is decreased or increased by means of the air pump (Boyle 1999, Vol. I, p. 210).

As CharlesWebster has stressed, some of the experiments in *Spring of the Air*, Experiments 7 and 17 for instance, were designed to investigate the quantitative relationship between the spring of the air and its degree of rarefaction.[8] Here Boyle did not meet with much success. However, within 2 years, Boyle had performed the experiment, oft repeated in school classrooms since then, involving a volume of air trapped by mercury in a U-tube the results of which supported the hypothesis that the spring of the air is proportional to its degree of rarefaction, that is, to its density.

I have not attempted to give a detailed and comprehensive account of Boyle's experiments in pneumatics in this place.[9] Rather, I have sought, by means of some examples, to illustrate how Boyle developed his concept of spring and deployed it to give mechanistic explanations of phenomena exhibited by his experiments.

---

[8] Webster (1965, pp. 467–470).

[9] Conant (1970) and Webster (1965) are classic studies of these developments and a new and controversial perspective on them is in Shapin and Schaffer (1985).

8.5 The Concept of Pressure in Boyle's Pneumatics

If Boyle is to be read as making significant advances in the treatment of fluids by introducing a concept of pressure then it is necessary to identify the way in which his version of the concept moved beyond the common sense that dates from antiquity without uncritically presuming that it involved all that is implied by the modern one. In my view, Boyle's writings on pneumatics, at least those prior to his reading of Pascal's Treatises in 1664, do not involve a technical sense of 'pressure' but only a common sense. As the title of his pioneering work suggests, the theoretical merit of *New Experiments Physico-Mechanical Touching the Spring of the Air and Its Effects* lay in the detailed way in which the spring of air, as distinct from its weight, was specified and brought to bear on the explanation of a range of pneumatic phenomena. Boyle did build on the work of predecessors like Torricelli, Roberval and Pecquet, to develop a technical sense of spring that he used to capture the distinctive characteristic of air that distinguishes it from solids. As a result of this spring, air presses on any surface with which it is in contact, such as the surface of the earth or the walls of a bladder. I claim that Boyle used the term 'pressure' in its common sense to refer to such instances of pressing. The only way in which the usage might be said to have moved beyond common sense was the explicit recognition that the force on a surface due to a sample of air under a given degree of compression is independent of the orientation of the surface and that the specification of that pressing as a force per unit area.

No formal definition or explication of 'pressure' is to be found in *Spring of the Air* which contrasts with the pains to which Boyle went in order to clarify the notion of 'spring'. Boyle's usage of the term without fanfare is already indicative of the fact that he considered himself to be employing a common concept in the usual way. The first occurrence of the term 'pressure' occurs in Boyle's elaboration of Experiment 1, which is concerned with the force on the sucker that is involved in the extrusion of air from the receiver of the air pump. When the handle attached to the sucker is released after such an extrusion the sucker is drawn up towards the receiver. This happens because of 'the protrusion of the external air, which, being much less rarefied than that within, the cylinder, must have a more forcible pressure upon the sucker, than the internal is able to resist'. In the next paragraph Boyle begins his exposition of the spring at work here.

> For the more easy understanding of the experiments tryable [sic] by our engine, I thought it not superfluous, nor unreasonable in the recital of this first of them, to insinuate the notion by which it seems likely that most, if not all, of them will prove explicable. Your Lordship will easily suppose, that the notion I speak of is, that there is a spring, or elastical power in the air we live in. By which ἐλατηρ or spring in the air, that which I mean is this: That our common air either consists of or at least abounds with, parts of such a nature, that in case they be bent or compressed by the weight of the incumbent part of the atmosphere, or by any other body, they do endeavor, as much as in them lies, to free themselves from that pressure, by bearing against the contiguous bodies that keep them bent (Boyle 1999, Vol.I, p. 165).

Note that it is weight or force of some other body that is the 'pressure' which compresses the air. This need involve nothing beyond the common sense of pressure. It is the spring of the air that is the novel 'notion' explicated by Boyle.[10] It is by virtue of this spring that atmospheric air presses any solid surface with which it comes in contact, that air trapped in a bladder can cause it to expand, that mercury is supported in a barometer and that prevents a smooth marble disc from falling from a second disc to which it is adjacent. The forces on the various surfaces involved in such cases are frequently described by Boyle utilizing a common concept of pressure that was far from novel.

Boyle moved beyond the common sense of his day by insisting and demonstrating that air has weight as well as a spring, although he was not the first to do so. The cause of pneumatic phenomena considered by rivals to be due to nature's abhorrence of a vacuum come about 'in consequence partly of the weight and fluidity of bodies here below and partly, perhaps principally, of the spring of the air, whose restless endeavor to expand itself every way, makes it either rush in itself, or compel the interposed bodies into all spaces, where it finds no greater resistance than it can surmount' (Boyle 1999, Vol. I, p. 245).

Pneumatic effects are a consequence of the spring, weight and fluidity of air. It is not difficult to understand why Boyle singles out spring as the principal cause. Because of its 'restless power to expand' air will fill any container into which it is put, its spring causing it to press against the sides of the container to a degree depending on its degree of compression. However, in the case of atmospheric pressure there is no containing vessel. In this case, there is a downwards force due to the weight of air, the lower layers of which are compressed as a result. At ground level atmospheric air is compressed to a degree that is dependent on the weight of the air above it, with a resulting equilibrium between the force due to the resulting spring, acting upwards on the air, and the total weight acting downwards. The air, being thus compressed, presses in all directions, on the ground beneath it and on the sides of anything that is immersed in it. It is the weight of the air that causes the compression at ground level, and the spring accompanying this compression that is responsible for the pressing. That the spring of the air is the principal cause of the pressing is illustrated by the fact that the same degree of pressing will ensue if the air is compressed to the same degree by some cause other than the weight of atmospheric air. As Boyle pointed out, this is illustrated by the fact that, in the lead up to his version of the 'void in the void' experiment, when a barometer is inserted into the receiver of the air pump prior to its evacuation, the level of the mercury remains unchanged from what it was before, in spite of the fact that in the receiver it is shielded from the atmosphere.[11] The spring of the air is the principal cause of

---

[10] Boyle frequently used the term 'notion' to describe the concepts involved in his science and stressed the fact that novel experimental advances make necessary the fashioning of novel notions or the modification of old ones. See, for example his remarks to this effect in his 'proemial essay', Boyle (1999, Vol. II, p. 20).

[11] As noted above, this kind of point had already been made by Torricelli in his response to Ricci's queries.

pressing. However, the weight of air needs to be included as a cause because without it the restless power of air to expand would ensure that there is no atmospheric air and no atmospheric pressure.

Spring and weight of air are on a different ontological level than the pressures to which they give rise. This point was made quite explicit by Boyle in *Continuation of New Experiments Physico-mechanical Touching the Spring and Weight of the Air, and their Effects* published a decade after the original study. The quotation is from Boyle's Experiment 11, in which he demonstrates that when the air is sucked out from the region above the mercury in a tube it rises in the tube until it reaches a maximum height if 29 inches or so. This was an effect Boyle attributed to atmospheric pressure, thereby removing the need to invoke nature's abhorrence of a vacuum. Boyle argued that

> whether there be or maybe a vacuum or not, there is no need to have recourse to a *fuga vacui* to explicate suction; and also whatever other causes have by *Gassendus* and *Cartesius* been ingeniously proposed to explicate the suction, it seems to depend clearly upon the weight of the atmosphere, or in some cases upon the Spring of the Air; though I deny not, that other causes may contribute to the pressure of the air; which I take to be the grand and immediate agent in these *phenomena* (Boyle 1999, Vol. VI, p. 62).

All pneumatic effects result from the pressing of air against solid surfaces. In that sense, pressure is the 'immediate' agent and it is a 'grand' one in the sense that it is involved in all such effects. The main, non-immediate, causes of these pressures are the weight and spring of air, although Boyle does not rule out the possibility that there may be others.[12] So pressures, that is, the pressings against a solid surface exerted by air, come about as a result of its weight or spring (to which pair of causes Boyle elsewhere adds fluidity, as we have seen) or some combination of them.[13]

Air, by virtue of the kind of thing that it is, possesses weight, fluidity and spring, the isotropic character of the spring being a consequence of the fluidity. As a consequence of these properties air presses against any surface with which it happens to be in contact. The common sense of 'pressure' can be readily utilized to describe such pressing, and this is precisely how Boyle used the term in his pneumatics, from

[12] On a number of occasions Boyle pointed to the phenomenon of capillary rise as an additional cause, significant when liquids in narrow tubes are involved.

[13] In *Leviathan and the Air Pump*, a book that, as observed by David Wootton (2015, p. 45), has been referred to as 'the most influential book in the [history of science] since Kuhn's *The Structure of Scientific Revolutions'*, Shapin and Schaffer find ambiguities in Boyle's use of 'pressure' that can be seen as of their own making once the relationship between spring and weight, on the one hand, and a common sense concept of 'pressure', on the other, is appreciated. According to our authors, Boyle used the term 'pressure' generically to refer to spring and weight. 'So "pressure" is to be read as an embracing term, and its ambiguities and variation of meaning were themselves a resource that Boyle used in debating the air-pump trials' (Shapin and Schaffer 1985, p. 55). By grouping together weight, spring and pressure as Boyle's 'principal ontological concern' Shapin and Schaffer interpret passages in Boyle as ambiguous and merely rhetorical which, on the account I offer, can be interpreted literally and not merely rhetorical as making clear and explicit claims about the relationship between pressure, in the common sense of the term employed at the time, and its causes, the spring, weight and fluidity of air.

the *Spring of the Air* and thereafter. This standpoint no longer proved to be adequate in the context of Boyle's treatment of pressure as exhibited by liquids.

8.6 Pressure in Boyle's Hydrostatics

Boyle had not focused on the issue of a theoretical characterization of liquids as distinct from solids and air in *Spring of the Air*. In that work Boyle distinguished *air* from solids by stressing that spring is distinct from weight, which solids and air both possess, and that the spring of the air is distinct from elasticity in solids through being isotropic and being expansive only. Forces are transmitted from point to point in mechanistic fashion in air on account of its spring. That is why Boyle had no need to develop a notion of pressure that went beyond the common sense one and which he could use to describe the forces exerted by air on solid surfaces. That standpoint could not work for liquids because they lack a corresponding spring, a fact that Boyle appreciated and had explored experimentally.[14] He seems to have become aware of the theoretical challenge posed by liquids, as distinct from solids and air, after reading Pascal's *Equilibrium of Liquids.* His response was spelt out in *Hydrostatical Paradoxes made out by New Experiments (for the Most Part Physical and Easy* (1666).

Apart from some minor qualifications, Boyle (1999 Vol. V, p. 206) accepted the theoretical content of Pascal's hydrostatics as 'worthy of the author'. However, he had problems with the experimental evidence Pascal invoked in its support. He raised doubts about whether Pascal had performed the experiments he described and whether they could possibly be performed following Pascal's prescriptions. His doubts ranged from the probability that stoppers in apparatus as described by Pascal would leak, to the observation that some of the experiments required that experimenters make extended observations twenty feet under water.[15] He also pointed out the difficulty of readily reproducing Pascal's experiments due to the fact that many of them required large quantities of mercury and glass tubes over twenty feet in length. Boyle (1999 Vol. V, p. 194) presented himself as an expositor of Pascal's hydrostatics seeking to clarify aspects of it by way of easy to perform experiments. He aimed to show how various hydrostatic phenomena, some of which can appear as 'surprising' and even 'abstruse', can be 'made out by new experiments (for the most part physical and easy)', as announced in the title of his tract.

Near the beginning of *Hydrostatical Paradoxes* Boyle characterized his objective as follows:

> It not being my present task to deliver the elements, or a body of hydrostatics, but only ten or twelve paradoxes, which I conceive to be provable by this new way of making them out, I shall to avoid confusion, deliver them in as many distinct propositions; after each of

[14] Boyle (1999, Vol. I, p. 168).

[15] Boyle (1999, Vol. V, pp. 206 and 255).

> which, I shall endeavor in a proof, or an explication, to show, both that it is true, and why it ought to be so (Boyle 1999, Vol. V, p. 207).

It is clear from his earlier remarks that Boyle aimed to go beyond 'geometricians' such as Stevin who, in his opinion, had asserted various propositions to be true without showing how or why they are true.[16] Boyle found the seeds of what he needed in this respect in Pascal's *Equilibrium of Liquids*. Pascal had indicated that hydrostatic forces are transmitted through liquids by virtue of their 'continuity and fluidity' and give rise to forces on bounding surfaces that are independent of their orientation.[17] However, Pascal did not specify the way in which forces are transmitted from point to point within the body of liquids. He did not explicate how it is possible for those forces to bend around corners. As I see it, Boyle's main contribution to the theoretical grasp of hydrostatics was a move in the direction of filling that gap.

Before embarking on his exposition in *Hydrostatical Paradoxes* Boyle (1999 Vol. V, p. 207) indicated that he would 'premise a word or two by way of either *postulation* or *lemma*'. I believe that his ensuing remarks reveal some innovative and consequential aspects of his approach, however innocuous they might appear to a modern eye on first reading. We are to consider a glass pipe, open at both ends, held vertically in a container of water with its lower end below the water surface. Boyle invokes 'an imaginary plane or surface, which touching that orifice [the lower end of the pipe] is parallel to the horizon; and consequently parallel as to sense to the upper surface of the water, and this being but a help to the imagination, will readily be granted'. Boyle proceeds to discuss various equilibrium conditions in terms of the forces acting on either side of this imaginary plane. The water will initially be at the same level inside and outside of the tube (provided the effect of capillary rise is negligible) because the water will press equally on all parts of the imaginary plane. If oil is poured on the water outside of the pipe the water inside the pipe will rise until the weight of the oil, acting on the plane outside of pipe via the main body of water, is countered by the additional water rising inside of the pipe. This strategy of considering the forces acting on an imaginary plane within a liquid, which Boyle introduces simply as an aid to the imagination, is of considerable consequence. By extending its use to situations more complicated than a vertical pipe partially immersed in water, including those involving forces on planes that are not horizontal, Boyle in effect traces how hydrostatic effects are transmitted from point to point through a liquid.

Boyle continued to use the term pressure in its common sense to refer to forces exerted by liquids on a solid surface. For instance, Paradox VI is concerned with the specification of the force exerted by liquids on the surfaces of immersed solids. Here we find Boyle specifying the 'pressure' on such a surface as equal to a column of liquid having a cross section equal to that of the surface in question and a height

[16] Boyle (1999, Vol. V, pp. 194–195).

[17] See Pascal (1937, pp. 7–8) for a clear and explicit expression of this point.

equal to the depth of that solid surface beneath the water surface.[18] However, Boyle extended his discussion to include forces acting laterally as well as vertically in the following terms:

> Whence we may learn (what I have not yet found mentioned by any writer,) that even in the midst of the water we may suppose a pillar of *water*, of a basis equal to the side of an immersed body, (and reaching to the lowest part of it;) and that, though this imaginary aqueous pillar – be not included in any solid body or stable superficies; nevertheless its lower parts will have a lateral pressure tending outwards, against the imaginary sides, from the weight of the water that is above these subjacent and lateral parts; and will have the pressure increased proportionably [sic.] to the height to which the imaginary pillar reaches above them (Boyle 1999, Vol. V, p. 248, emphasis in original).

Here Boyle explicitly uses the concept of pressure in the extended sense that I have identified, to refer to the action of water on water. In addition, his reference to the fact that such an extension is not to be found in the work of any other writer suggests that his own conception of his originality on this point coincides with mine. In Boyle's exposition of hydrostatics the reference of the term 'pressure' has been extended, from the forces on the boundaries between solids and liquids, to include the equal and opposite forces that act on each other across imaginary boundaries within the body of liquids in equilibrium.

The shift I have identified, from a common sense concept of 'pressure' to a more technical sense that serves to distinguish solids from fluids is exemplified particularly clearly in *An Hydrostatical Discourse*.[19] In that tract, published in 1672, Boyle responded to criticism of his position by Henry More, the general drift of which was that the former's pneumatics and hydrostatics needed to be bolstered by reference to non-mechanical causes such as the tendency of the elements to move to their natural places in order to adequately explain the relevant phenomena. In his reply, Boyle (1999 Vol. VII, p. 159) made it clear that he was more than ready to 'assert an incorporeal being that made and governs the world', but insisted that appeal to incorporeal agents were not necessary to accommodate pneumatics and hydrostatics since they could be explained 'mechanically'. In his subsequent defense of this position we find Boyle articulating and clarifying a *mechanistic* conception of 'pressure' as a cause distinct from weight.

In one specific interchange, More invoked an experiment which he saw as raising doubts about the extent to which hydrostatics can be explained mechanically. It involved the recognition that if a disc of wood, less dense than water, is forcibly held at the bottom of a container of water and then released, it rises to the surface. In his example, More supposed that the surface area of the wooden disc is only a little smaller than that of the cross sectional area of the vessel of water in which it is immersed. He pointed out that the weight of the wooden disc plus the weight of the column of water vertically above and pressing down on it is greater than the weight

[18] Boyle (1999, Vol. V, p.239. This much was entailed by Stevin's *Elements of Hydrostatics*, but, to use Boyle's words, Stevin had asserted that this was true without showing why it was true. The inadequacies of Stevin's proofs were discussed in Chap. 3.

[19] The specific interchange on which I focus is in Boyle (1999, Vol. VII, pp. 158–164).

of water surrounding the disc. More concluded that considerations of weight as understood by Boyle are incapable of explaining why the disc rises to the surface.[20]

In his reply, Boyle in effect agreed with this latter claim, but argued that the problem is solved, not by assuming, as More did, that water does not weigh in its 'proper place', nor by invoking any other non-mechanical cause, but by invoking pressure, a mechanical cause distinct from weight. I will not rehearse the details of the interchange in this place because the issue at stake is brought out more clearly by an ingenious experiment of Boyle's own that he invoked. A focus on that experiment also enables me to draw an illuminating comparison with Galileo's deployment of a very similar experiment, one that was discussed in Sect. 4.3.

In the experiment in question Boyle poured some molten wax into a flask whose base was covered with a shallow layer of water. When the wax cooled and solidified it contracted into a solid body of wax leaving a small space between it and the walls of the containing vessel. When water was introduced into this small space the wax, being slightly less dense than water, was raised so that its upper surface remained a little above the water surface. The rising of the wax by the water took place in spite of the fact that 'this collateral water was so far from being heavier than the wax its pressure impelled up, that both the collateral and the stagnant water all together, being weighed in good scales, amounted to little above a quarter of the wax'.[21] Boyle (1999 Vol. VII, p. 161) insists that here, as in More's experiment and in the case of immersed bodies in general, the phenomenon is explained by invoking 'the pressure of the ambient fluids on immersed solids as uniform or every way equal'. The wax floats because the force arising from the pressure on its lower surface exceeds that arising from the pressure on its upper surface by an amount sufficient to counteract its weight.

The notion that floating bodies experience an upwards thrust equal to the weight of water they displace encourages an understanding of floating involving balancing weights, an interpretation to which Archimedes' original text on floating bodies lends itself.[22] A solid displaces an amount of water that weighs down as a consequence of its propensity to return to its former place and floating occurs when that weight balances the weight of the immersed solid. Boyle's experiment undermines this interpretation as did More's, since, in the circumstances that they each consider, the displaced water weighs much less than the floating body. Galileo had adopted and refined the account of floating in terms of balancing weights over half a century

[20] An English translation of the relevant passages from More's *Enchiridium Metphysicum* of 1671 is given in Boyle (1999, Vol. VII, p. 160). For a complete English translation of More's *Enchiridium Metaphysicum* see Jacob (1995). Chapters XII and XIII, the object of Boyle's critique, are in Part II, pp. 19–102 of that work.

[21] Boyle (1999, Vol. VII, pp. 162–163).

[22] Proposition 5 of Book 1 of Archimedes' *On Floating Bodies* reads, 'Any solid lighter than a fluid, if placed in the fluid, be so far immersed that the weight of the solid will be equal to the fluid displaced' which is not the case in Boyle's experiment with the wax. See Heath (1950, p. 257).

earlier, as was discussed in Chap. 4. It is instructive to compare his treatment with that of Boyle.

Galileo's explanation of the phenomenon differs markedly from that of Boyle. Galileo dealt with it by invoking the principle that when balancing weights are slightly displaced from their equilibrium positions in a vertical direction the ratio of those weights is inversely proportional to the ratio of their respective velocities. Galileo had illustrated his position by reference to a block of wood floating in water in a vessel into which it barely fitted, leaving room for very little water. A slow depression of the wood causes a small amount of water to rise quickly around it. Consequently, 'a very small quantity of water may raise up and sustain with its small weight a solid body that is a hundred or a thousand times heavier'.[23] The situation is remarkably similar to Boyle's experiment with wax.[24] In this way Galileo considered himself to have identified the 'true cause' of floating (Drake 1981, p.41).

I have sympathy with the view, already voiced by Stevin, that hypothetical displacements that do not in fact take place when systems are in equilibrium cannot function as the true cause of that equilibrium.[25] Once equilibrium conditions are known, they can be shown to be the correct ones by applying Galileo's principle. But, certainly if one is concerned with mechanistic explanations, this falls short of identifying the causes of floating. Such explanations were identified by Boyle, and he did so by invoking pressure. The phenomena figuring in the above discussion 'depend upon a mechanical aequipollence of pressure'.

> By this you may see that for the regulation of hydrostatical things, Nature has her balance too as well as Art, and that in the balance of nature the statical laws are nicely enough observed (Boyle 1999, Vol. VII, p. 164).

The balance referred to by Boyle here is due to forces arising from pressure as well as weight.

---

[23] Drake (1981 p. 26). A detailed discussion of these views of Galileo and the path that led to them can be found in Palmieri (2005)

[24] There is no direct evidence that Boyle was drawing on Galileo's work here. He did, in Boyle (1999, Vol. V, p. 194), cite Galileo among those of his predecessors who made considerable contributions to hydrostatics. However, Boyle included him amongst those who handled hydrostatics 'rather as geometricians than as philosophers', a judgment that makes sense in light of the contrast between the treatment of floating by Boyle and Galileo that I am highlighting here.

[25] Stevin's critique of the appeal to imaginary displacements as causes can be found in Dijksterhuis (1955, p. 509). However, he did not respond to the problem by supplying mechanistic causes as Boyle was able to do.

8.7 The Novelty of Boyle's Concept of Pressure in Liquids

In my account of Boyle's development of the concept of pressure in the context of liquids I have identified two ways in which his contributions were novel. They are his conception of pressure as acting on either sides of imaginary planes in the body of fluids and the extent to which he separated weight and pressure as distinct causes of the behavior of liquids. The first of these contributions was anticipated by Stevin to some degree and both of them were likewise anticipated by Pascal. In this section I discuss those anticipations and refine my attribution of novelty to Boyle in light of them.

As we saw in Chap. 3, Stevin introduced two technical terms in his *Elements of Hydrostatics* that to some degree resemble Boyle's imaginary planes. According to Stevin's Definitions VII and VIII a 'surface vessel' is 'the complete geometrical surface of a body, conceived as separable therefrom' and 'bottom' [*bodem*] is 'any plane against which rests any water'. Stevin explicitly likened the former to the planes of geometry, describing them as 'vessels without any corporeal magnitude and without any weight' (Dijksterhuis 1955, p. 385). Although they do not have 'corporeal magnitude' Stevin's 'surface vessels' are solid insofar as they can contain water, in accordance with the nature of solids as specified in Definition VI that reads 'Solid body is one whose matter does not flow, and through which penetrates neither water nor air'. The surface vessels need to be solid surfaces in order for him to utilize his Postulate III, the only postulate that introduced significant hydrostatic content into his theory. 'The weight causing a vessel to sink less deep to be lighter, but the weight causing it to sink deeper to be heavier, and that causing it to sink to the same depth, equally heavy.' This postulate is one that can be 'granted' insofar as it is an abstraction from common experience of the effect of adding or subtracting weights to floating vessels, which need to be solid to hold or support them. It is also clear from the context in which Stevin uses the term 'bottom' that the surfaces so named, although lacking thickness and weight, are *solid* surfaces against which water can press. The solid nature of 'bottoms' is made quite explicit when Stevin describes those bordering a rectangular prism of water as 'a corporeal rectangle' (Dijksterhuis 1955, p. 415). So, in spite of the degree to which Stevin's surface vessels and bottoms abstract from weight and corporeal magnitude, they do not abstract from solidity as possessed by solids and so cannot straightforwardly perform the function played by Boyle's imaginary planes against which liquids press in a way not involving pressure against a solid surface. It is very easy to read the modern conception of pressure and Boyle's moves towards it into Stevin's writings. Dijksterhuis does this to a certain extent through his extensive use of the term 'pressure' in his translation of a variety of terms in Stevin's text, terms that were clearly common sense rather than technical ones.

There is one place at least where Stevin did state that a body of water presses on an adjacent body across the plane that separates them. He did so in *The Practice of Hydrostatics* when discussing a situation where two columns of liquid of equal height but unequal volumes press against each other. As my discussion in Sect. 3.9

shows, Stevin began by imagining the two columns to be separated by a 'bottom', represented by CD in Fig. 3.3, p. 45, and considered the question of whether the different amounts of water pressed equally on either side of the bottom. He concluded that they did so from the fact that the water levels would remain equal were the bottom, CD, to be removed. 'Let the bottom CD be taken away; then the large water will thrust against the smaller in that place' (Dijksterhuis 1955, p. 490).

To a large degree, the idea that within a mass of water, adjacent bodies of water press on each other is a matter of common sense. Boyle drew upon that common sense when he specified a more formal sense of the way in which a liquid presses adjacent bodies of liquid across the imaginary planes, not solid planes, separating them. Boyle stressed both the innocuousness of his move, representing it as a mere aid to the imagination, and his originality, by describing the move as one that had not been made by any other writer.

In Sect. 6.6 I pointed to pages in Pascal's *Equilibrium of Liquids* where instances of hydrostatic effects being brought about by water pressing on water are at least implied. I also noted that Pascal went some way towards separating weight and pressure when he distinguished between forces arising from an external force applied at some location of the surface of a liquid and forces arising from the weight of the liquid transmitting those forces. In neither case did Pascal spell out the implications that such realizations had for an understanding of pressure as acting within the bodies of liquid in the way that Boyle was to move towards.

8.8 Further Comparisons of the Hydrostatics of Boyle and Pascal

As I pointed out early in the previous section, Boyle was motivated to give serious attention to the formulation of hydrostatics by Pascal's *Equilibrium of Liquids.* The style of reasoning at work in Pascal's treatise was 'experimental' in contrast to the Euclidean style of Stevin's hydrostatics. Whereas the content of Stevin's hydrostatics was to be derived from postulates sufficiently unproblematic to be granted at the outset, Pascal's version was to be rendered intelligible and grantable by demonstrating the range of phenomena, including phenomena elicited by experiment, which could be readily and straightforwardly explained by appeal to it.[26] Boyle followed Pascal in this respect and improved on the latter's efforts insofar as he applied the new hydrostatics to a range of straightforward experiments capable of informing and being performed by 'persons no more than moderately versed in the vulgar principles of hydrostatics' and, moreover, ones that Boyle did in fact perform (Boyle 1999, Vol. V, p. 207). Boyle solidified the case for Pascal's hydrostatics by improving the quality and quantity of its experimental basis. This point notwithstanding, my focus in this Section is on the relationship between the theoretical contributions

[26] This comparison is developed in some detail in Chalmers (2015a).

to hydrostatics made by Pascal and Boyle. There is an important sense in which the two contributions were complementary; Boyle supplied an account of the mechanisms responsible for communicating hydrostatic forces through the body of liquids and Pascal stressed the extent to which hydrostatics could be accommodated under the umbrella of general mechanical principles.

The approaches of Pascal and of Boyle have been differentiated by describing the former as mathematical and the latter as experimental.[27] Such a contrast is not borne out by the texts with which we have been concerned. Apart from the fact that Pascal provided and stressed the experimental basis of his theory, neither Pascal's *Equilibrium of Liquids* nor Boyle's *Hydrostatical Paradoxes* abound with geometrical demonstrations as is the case with Stevin's *Elements of Hydrostatics.* The former theories are mathematical to the extent that hydrostatic forces on surfaces need to be expressed and treated as forces per unit area, but they are alike in that respect. Notwithstanding this similarity between the treatments of hydrostatics by Pascal and Boyle, there is a theoretical emphasis in Pascal treatment that is absent from Boyle's.

Pascal's theoretical grasp of hydrostatics was expressed in his appreciation of the fact that devices such as the hydraulic press can be understood as 'machines for multiplying force' analogous to the simple machines figuring in the science of weight. As such, the former could be seen as conforming to the same principles as the latter. With respect to the hydraulic press Pascal (1937 p. 6) wrote:

> It is remarkable that this new machine exhibits the same constant relation that is characteristic of all the old machines, such as the lever, the wheel and axle, the endless screw, and others, which is that the distance traversed increases in the same proportion as the force.

Pascal extended the analogy with simple machines by noting that 'it is evident that it amounts to the same thing whether we make one hundred pounds of water move through an inch, or make one pound of water move through one hundred inches'. He also noted that the press also conforms to the principal that 'a body never moves by its own weight without lowering its center of gravity'.[28] Further, Pascal invoked the hydraulic press to bring out in a forceful way the distinction between the behavior of solids and liquids. Since the force is transmitted from one aperture of the hydraulic press to the other by virtue of the 'continuity and fluidity' of the water, the functioning of the press is destroyed if the communicating water is frozen. Since the freezing does not affect the continuity, it is clearly the fluidity possessed by the water and not by the ice that is crucial for explaining the hydraulic press and other hydrostatic phenomena.

None of the points figuring in the previous paragraph appear in Boyle's *Hydrostatic Paradoxes*. The hydraulic press does not get a mention. Boyle was concerned, not to embrace hydrostatics within the scope of some general mechanical

---

[27] See the recent discussion of this issue in Malet (2013) and the references cited there.

[28] Pascal (1937, p. 8). As noted in Chap. 6, Pascal introduced this point by announcing it as a 'proof which will be understood only by geometers and may be disregarded by others', thereby reinforcing my rejection of the characterization of Pascal's approach as mathematical as opposed to experimental.

principles, but rather to identify how forces are transmitted from one location to another within liquids and thereby bringing about hydrostatic effects. The utility of Boyle's stance can be brought out by the following reflection. In the science of weight, the application of the principles appealed to by Pascal to particular instances requires a specification of the constraints at work. In the case of simple machines, such as the balance for example, the constraints are obvious and taken for granted. It is the rigidity of the balance arm that relates the distances moved by a weight to that moved by the one it balances. When it comes to hydrostatic machines the precise nature of the constraints are not so obvious. Boyle's moves towards the identification of the way in which forces are transmitted through liquids supplies what is needed here in a more explicit way than is accomplished in Pascal's *Equilibrium of Liquids*.

Pascal had identified fluidity as essential to the behavior of liquids. Because of this, together with the continuity of liquids, a force applied to an area at one point in a liquid will be transmitted throughout the liquid, exerting the same force per unit area on any other boundary of the liquid, however remote and whatever its orientation.[29] This falls short of tracing the way in which forces are transmitted across boundaries in the body of the liquid. It falls short of using 'pressure' to identify the force per unit area acting between neighboring portions of liquid through the body of that liquid, as Boyle came to do. Pascal talked freely of water pressing against apertures using the verb '*presser*' but did not talk of pressure within the body of liquids. Pascal did not use the noun '*la pression*' in *Equilibrium of Liquids* or *Weight of the Air*. It was Boyle who furthered the cause of giving a mechanistic account of hydrostatic phenomena by invoking the forces acting on imaginary planes within the body of a liquid and thereby interpreting pressure as acting in such locations. In this way he set the scene for Newton to develop a theory that would explain mechanistically how hydrostatic forces can bend around corners.

I conclude this comparison of Pascal and Boyle on hydrostatics with a reflection on the fact that, unlike Pascal, Boyle undoubtedly performed the experiments that he invoked. Near the beginning of *Hydrostatical Paradoxes* Boyle (1999 Vol. V, pp. 206–7) announced that he sought to explicate hydrostatics 'with so much more ease and clearness, that not only This Illustrious Society [The Royal Society] but persons no more than moderately versed in the vulgar principles of hydrostatics, may easily enough apprehend what is designed to be delivered'. Now it is one thing to devise a means of conveying an idea, such as Boyle's novel conception of pressure, to the reader. It is another to arrive at the idea in the first place. The understanding of pressure as acting on imaginary surfaces within the body of liquids was original with Boyle. I suggest that, in the course of his probing with his search tube, Boyle came to think of himself as detecting the degree of pressing at various locations and in various directions within a liquid by means of it. Einstein may have thought with his pen. Here we find Boyle thinking with his search tube. I conjecture that the new conception of pressure arose as a result of Boyle's reflection on extensive material manipulations that he actually performed.

[29] See Pascal (1937, pp. 7–8) for a clear and explicit expression of this point.

8.9 Pressure and Fluidity as Novel, Intermediate, Mechanistic Causes

Boyle was a leading proponent of the movement that by the 1660s was being referred to by him and others as the experimental philosophy. Pneumatics and hydrostatics exemplified the new kind of science insofar as it was to be rendered intelligible by reference to, and supported by, experiments rather than via Euclidean-style proofs from given premises. As far as pneumatics was concerned, experiment was also essential insofar as it was only by way of it that many of the key phenomena became apparent. Torricelli's and the Puy de Dôme experiment provide ready examples and Boyle was able to expand the range of novel experimental effects by way of his air pump. By the time Pascal and Boyle turned their attention to a theoretical formulation of hydrostatics, a need for novel experimenting was not pressing since a wide range of relevant hydrostatic effects were already familiar, as is evident, for instance, from Marin Mersenne's survey and extension of relevant phenomena in 1644.[30] A strong case for the theory in Pascal's *Equilibrium of Liquids* could be made by appealing to the natural way in which it could account for a range of well-known effects, which is why he could be confident of the results of further experiments that he had not in fact performed.[31]

Boyle shared with many of his contemporaries the desire to rid philosophy of what had come to be seen as the obscure appeal to Aristotelian forms and aimed to replace them with contact action between particles of matter. As far as the account of the ultimate ontology of the material world figuring in Boyle's mechanical philosophy is concerned, it involved the shapes, sizes and motions of particles of universal matter and nothing else. When it came to the less ambitious, but for Boyle very urgent, task of establishing explanations of particular kinds of observable phenomena that were accessible to experiment he took his cue from the ancient science of simple machines. He sought to move beyond that science by identifying causes in addition to weight capable of affording the kinds of mechanistic explanations that it involved in fresh domains. Like weight, the new causes were to be experimentally accessible and yet intermediate insofar as they lacked an explanation in terms of the ultimate ontology of the mechanical philosophy. As far as pneumatics and hydrostatics were concerned, the novel, intermediate causes elaborated on by Boyle were 'spring' and 'fluidity', the latter being elucidated by appeal to a concept of pressure.

In pneumatics the relevant phenomena were linked to their causes by way of the spring of the air. The relevant pushes were transmitted form point to point through a body of air in a way analogous to that in which a force is transmitted from point

[30] See Mersenne (1644, pp. 215–233). As noted in Duhem, (1905), Mersenne's somewhat disorganized survey of the hydrostatics available to him contains some original observations. They include an anticipation of the hydraulic press (Mersenne, 1644, p. 228) and the recognition that hydrostatic effects are destroyed if the water is frozen (Mersenne, 1644, pp. 228 and. 229). The latter page is wrongly numbered as 239 in the copy of Mersenne (1644) available on http://books.google.com.

[31] This position is argued in Chalmers (2015a).

to point through a stretched elastic string. This conception will not do in hydrostatics since liquids lack a significant 'spring'. Following his reading of Pascal's hydrostatics, and building on the emphasis to be found there on the fluidity of liquids as crucially responsible for the transmission of hydrostatic forces, Boyle came to understand fluidity in terms of pressure acting isotropically throughout the body of liquids. He used the device of considering the forces acting on either side of imaginary planes within the body of a liquid. This was taken for granted by Newton when he turned his attention to hydrostatics and was a key factor in accounting for his success. Newton's hydrostatics is the topic of the next chapter.

References

Boyle, R. 1990. *Collections from the Royal Society: Letters and papers of Robert Boyle*. Bethesda: University Publications of America.

———. 1999. In *The works of Robert Boyle*, ed. M. Hunter and E. Davis, vol. 14. London: Pickering and Chatto.

Chalmers, A. 2015a. Qualitative novelty in seventeenth-century science: Hydrostatics from Stevin to Pascal. *Studies in History and Philosophy of Science*. 51: 1–10.

———. 2015b. Robert Boyle's mechanical account of hydrostatics and pneumatics: Fluidity, the spring of the air and their relationship to the concept of pressure. *Archive for History of Exact Sciences*. 69: 429–454.

Conant, J.B. 1970. Robert Boyle's experiments in pneumatics. In *Harvard case studies in experimental science*, vol. 1, 1–63. Cambridge, MA: Harvard University Press.

Dijksterhuis, E.J. 1955. *The principal works of Simon Stevin, Volume 1, Mechanics*. Amsterdam: Swets and Zeitlinger.

Drake, S. 1981. *Cause, experiment and science: A Galilean dialogue incorporating a new English translation of 'Bodies that stay atop water or move in it'*. Chicago: University of Chicago Press.

Duhem, P. 1905. Le Principe de Pascal. *Revue Générale des Sciences Pures et Applique*. 16: 599–610.

Heath, T.L. 1950. *The works of Archimedes*. New York: Dover.

Hunter, Michael. 1995. How Boyle became a scientist. *History of Science*. 33: 59–103.

Jacob, Alexander. 1995. *Henry More's manual of metaphysics: A translation of the Enchiridium Metaphysicum. Parts I and II*, Studien Materialien zur Geschichte der Philosophie. Vol. 38. Hildesheim: G. Olms Verlag.

Malet, Antoni. 2013. Between mathematics and experimental philosophy: Hydrostatics in Scotland in 1700. In *The Mechanization of natural philosophy*, ed. D. Garber and S. Roux, 159–187. Dordrecht: Springer.

Mersenne, M. 1644. *Cogitata Physico-mathematica*. Paris: Antoni Bertier.

Newman, W. 2003. *Gehennical fire: The life of George Strarkey, an American alchemist in the Scientific Revolution*. Chicago: University of Chicago Press.

Newman, W., and L. Principe. 2002. *Alchemy tried in the fire: Starkey, Boyle and the fate of Hermontian chymistry*. Chicago: University of Chicago Press.

Pascal, B. 1937. *The physical treatises of Pascal: The equilibrium of liquids and the weight of the mass of the air*. Trans. A. G. H. Spiers and I. H. B. Spiers. New York: Columbia University Press.

Shapin, S., and S. Schaffer. 1985. *Leviathan and the air pump: Hobbes, Boyle and the experimental life*. Princeton: Princeton University Press.

Webster, C. 1965. The discovery of Boyle's law and the concept of the elasticity of air in the seventeenth century. *Archive for History of Exact Sciences*. 2: 441–499.

Wootton, D. 2015. *The invention of science: A new history of the scientific revolution*. London: Penguin Random House.

Chapter 9
Newton's Hydrostatics: Liquids as Continua

Abstract Following an early version in an unpublished manuscript, Newton included a concise version of his hydrostatics in the *Principia.* It was based on his identification of the property of liquids that distinguishes them from solids, namely, the inability of the former to resist distorting forces. Any portion of a liquid, however small, is presumed to possess this property. In this way Newton made precise theoretical sense of the continuity and fluidity of liquids that had been presupposed in a more common sense by Stevin, Pascal and Boyle before him. Newton began his treatment of hydrostatics by considering a sphere of non-gravitating liquid. In doing so, he clearly distinguished pressure from weight. He freely adopted the technique, introduced by Boyle, of considering the pressures acting on either side of planes within the body of a liquid. The strict continuity of liquids enabled Newton to construct proofs that relied in spherical surfaces touching planes or other spherical surfaces at points. Newton was able to show how his definition of a liquid has the consequence that, within a body of liquid in equilibrium a pressure exerted on its bounding surface is transmitted as the same force per unit area normal to any surface within the body of liquid, whatever its orientation. Having achieved that result, Newton then introduced gravity and showed how the pressure in a liquid is proportional to the depth below its surface. Newton construed the unpublished version of his theory as a mathematical one following from grantable assumptions. It is better seen as an empirical theory, with its capacity to explain a range of observable and experimental phenomena providing the evidence justifying his definition of the liquid state.

9.1 Situating Newton's Hydrostatics

As is well appreciated, and is evident from the entries in Isaac Newton's early notebooks, his early engagement with mechanics grew out of a close study of the work of Rene Descartes, especially the mechanical system of the world presented in the latter's *Principles of Philosophy*. Newton's early treatment of particle collisions and of circular motion, appearing in the *Waste Book* from the mid-1660s, can be seen as a deployment and modification of Descartes' account of the inertial motion and the centrifugal forces that, for example, the latter presumed to be experienced by a

A.F. Chalmers, *One Hundred Years of Pressure*, Archimedes 51,
DOI 10.1007/978-3-319-56529-3_9

particle rotating at the end of a sling.[1] Newton did not need to rely on Descartes to provide him with a starting point for his hydrostatics. As we have seen, a tradition was available elsewhere for him to draw on, especially in the work of Simon Stevin, Blaise Pascal and Newton's contemporary and early inspiration, Robert Boyle. The articulation of hydrostatics by those authors assumed liquids to be continuous, a theme that I elaborated on below. Eventually the new hydrostatics involved the addition of pressure to weight as the causes of hydrostatic effects. The tradition in question did not rely on any explanation of those causes by appealing to underlying corpuscles, and differed from Descartes' approach in that respect. The hydrostatic tradition, as I hereafter refer to the developments to which Stevin, Pascal and Boyle made key contributions, constituted a fruitful starting point for Newton's venture into the area in a way in which Descartes' corpuscular hydrostatics did not.

While Newton could not have found positive guidance for his hydrostatic in Descartes' *Principle of Philosophy* there is a sense in which it provided him with a positive reason to turn his attention to the development of the hydrostatic tradition in its stead. In the notebook *Certain Philosophical Questions*, composed in the early to mid-1660s Newton had already noted that Descartes' conception of light clashed with the linear transmission of the latter. The line of action of one spherical corpuscle pushing on another, the mechanism by which light is transmitted according to Descartes, is only transmitted linearly in the special case that the adjacent particles are arranged in a straight line. Once the corpuscular picture is set aside and a theory in which the full isotropy of hydrostatic forces is recognized then the incompatibility of the understanding of light as a pressure and the linear transmission of light rays becomes even more pronounced.[2] It may well have been the case, then, that Newton was motivated to include hydrostatics into his mechanics by the fact that such a theory would make possible a definitive rebuttal of Descartes' theory of light. This made way for his own version that he struggled to develop and refine for a large part of his working life.

Central to and distinctive of Newton's mechanics, from the early efforts recorded in his notebooks, especially the *Waste Book,* up to the fully worked out dynamics of the *Principia*, was his deployment of a technical notion of force as a cause of change in motion. As far as statics is concerned, equilibrium was to be understood in terms of balancing forces. Accordingly, hydrostatic equilibrium was to be comprehended in terms of the distinctive way in which forces act in liquids, as distinct from the way they act in solids. The aim was to precisely, that is, mathematically, define the essential nature of liquids in this respect in such a way that the propositions of hydrostatics could be derived from that definition in conjunction with the principles of mechanics. That this was indeed Newton's goal is clear both from his first venture into hydrostatics recorded in the work usually referred to as *De gravitatione,* the opening words of Newton's text, and from the later version appearing in the

[1] I have referred to the version of the *Waste Book* in Herivel (1965, pp. 128–182). The same work (pp. 1–34) contains a discussion of Newton's early moves in dynamics.

[2] The incompatibility of the construal of light as a pressure and its linear transmission is explored in great detail in Shapiro (1974).

Principia. Newton did not extend the range of phenomena explained within the hydrostatic tradition. Rather, he aimed to deepen an understanding of it by spelling out its theoretical structure and foundations. As we shall see, Newton very nearly achieved his goal.

9.2 The Two Versions of Newton's Hydrostatics

The first version of Newton's hydrostatics appears in the unpublished tract *De gravitatione* in the first sentence of which Newton made it clear that his topic was 'the science of gravity and the equilibrium of fluids and solid bodies in fluids'. In the remainder of the opening paragraph Newton distinguished two methods for treating the topic. The first of them involves the demonstrations of the propositions of hydrostatics 'from abstract principles, sufficiently well known to the student, strictly and geometrically'. The second method involves strengthening the case for the propositions and demonstrating their usefulness by appeal to experiment. For Newton, in the mathematical sciences it is essential to abstract from physical considerations, so it is the first method that has priority. 'I shall not be reluctant to illustrate the propositions abundantly from experiments as well [as geometrical demonstrations], in such a way, however, that this freer method of discussion, disposed in scholia, may not be confused with the former, which is treated in Lemmas, propositions and corollaries.' Newton's aim was to present hydrostatics as a body of propositions geometrically demonstrated from abstract principles, the latter being unproblematic insofar as they are 'sufficiently well known to the student'.[3]

Within one page of having embarked on his task Newton got caught up in a lengthy digression in which he offered a detailed critique of Descartes' philosophy of space, matter and motion and offered his own alternative to it. It is the content of the digression that, understandably, has attracted the attention of scholars to *De gravitatione*, but it is not my concern here. Three quarters of the way into the article Newton admits that he has 'digressed enough' and returns to the project outlined in his opening paragraph. The text ends somewhat abruptly before the project is completed, but it is sufficient to constitute a significant contribution to the theorization of hydrostatics. My main concern is with the content of the concluding few pages of *De gravitatione* in which Newton's first version of hydrostatics is presented.

There is considerable disagreement in the literature concerning the date of composition of *De gravitatione*. Hall and Hall (1962, especially pp. 89–90) made a case for an early publication date, around 1665–1668, based largely on the extent to which the argument of *De gravitatione* involves a detailed engagement with the views expressed in Descartes' *Principles of* Philosophy, characteristic of the concerns of a young Newton as portrayed in his early notebooks. A date of composition around that time was generally accepted until it was challenged in the 1980's by

[3] I have used the English translation of *De gravitatione* in Hall and Hall (1962, pp. 121–156). The opening paragraph discussed here is on p. 121.

Betty Dobbs.[4] Dobbs finds in *De gravitatione* echoes of themes found in Newton's *Principia* and suggests a date of around 1884/5, only 2 or 3 years before the publication of Newton's masterpiece. Important for Dobbs' case is the extent to which Newton's discussion of the resistance offered to the motion of comets and pendulum bobs by the ether can be taken as presuming knowledge of discussions following the return of Halley's comet in 1982 and also of new experiments with pendulums.[5] Those discussions and experiments were recorded in a series of documents by Newton under the title *De Motu,* composed in 1684/5. Dobbs argues that *De gravitatione* must have been written around that time. More recently Steffen Ducheyne (2012, pp. 129–131) has argued for a composition date of 1683/4 along similar lines. John Henry (2011), basing his argument on how the discussion of *De gravitatione* fits in with Newton's changing views on action at a distance, opposes Dobbs' case and argues for a return to 1668 as the most likely date of composition, in keeping with the estimate of the Halls. J. A. Ruffner (2012) presents a circumspect review of the state of the discussion.

In light of the analysis of Newton's hydrostatics that follows in this chapter, the account of it in *De gravitatione* must be seen as an extension of and improvement on the hydrostatic tradition represented in the work of Stevin, Pascal and Boyle. That is the case whatever the date of its composition. It is highly likely that Newton became aware of the hydrostatic tradition via Boyle's *Hydrostatical Paradoxes*, published in 1666. For this and other reasons, to be discussed in Sect. 9.9, I favor a date of 1668 or soon thereafter as the most likely date of composition. However, I repeat, my analysis of the structure of the argument of the hydrostatics in *De gravitatione* and its relation to the hydrostatic tradition that preceded it is independent of that preference.

The hydrostatics of *De gravitatione* was followed by a second version that appeared in the *Principia.*[6] Having spelt out the dynamics of the motion of bodies in Book I of that work, Newton turned his attention, in Book II, to the motion of bodies in resisting media. In the first four Sections he explored in mathematical detail the motions in fluids whose resistance is characterized by various idealized force-laws. At the conclusion of Section IV he indicated his intention to focus on the nature of the media through which solid bodies can move. The result was Section V, headed 'The density and compression of fluids and hydrostatics'. In a few pages of concise text Newton spelt out; an account of pressure then an account of the pressures arising from gravitation. The account of pressure differs from that appearing in *De gravitatione* by giving a clearer definition of a fluid and deriving the isotropy of pressure from it via a valid proof. The extension, in the second part, to include

---

[4] See Dobbs (1991). Alan Gabbey (2011, note 19) has fixed 1668 as the earliest possible date of composition based on a reference by Newton to a source he could not have had access to prior to that date.

[5] Newton's discussion of comets and pendulums is in the final paragraph of Newton's digression on space, time and motion in *De gravitatione*.

[6] All of my quotations from the *Principia* are from the translation by I.B. Cohen and A. Whitman in Newton (1999). Newton's concise version of hydrostatics is on pp. 687–692.

gravitation, went beyond what was in *De gravitatione* and plausibly constitutes the topic that Newton would have covered had he completed it.

The version in the *Principia* is more concise and precise than the *De gravitatione* version. On the other hand, the earlier text is more informative insofar as, in it, Newton discussed the nature of his hydrostatic project at a meta-level, as it were, raising explicitly questions about the method he was employing and the assumption that fluids are continuous.

9.3 *De gravitatione*: The Background

Scholars have paid little attention to Newton's hydrostatics. A notable exception is a seminal article by Alan Shapiro (1974) in which he compared Newton's hydrostatics with that of Descartes, tracing the way in which Newton's grew out of an engagement with and critique of Descartes'. Shapiro's discussion necessarily involved consideration of the connections between hydrostatics and the theory of light because in Descartes' view the transmission of light through the ether and that of gravitational effects through fluids were virtually identical processes, as was discussed in Chap. 5. As I mentioned in my introduction to this Chapter, Newton may well have been motivated to develop a theory of hydrostatics because in the light of such a theory Descartes' account of light could be refuted.[7] In the following I make much use of Shapiro's account of all this, adapting it to my own purpose. However, I aim to complement it in an important respect. I believe that Shapiro's emphasis on the Descartes/Newton connection overlooks another important source, namely, the hydrostatic tradition. What Newton could get from that tradition that he could not get from the writings of Descartes was an account of hydrostatics that could explain a range of otherwise problematic phenomena and which was based on the presumption that fluids are continuous.

In passages in his *Waste Book*, entries into which began around 1664, we find Newton making steps towards a new dynamics by way of an analysis and critique of Descartes' *Principles of Philosophy*, to which Newton explicitly referred. Newton was innovative in his analysis of collisions and in his analysis of circular motion. In both cases the focus is on particles, colliding spheres and a sphere moving on the interior surface of a spherical surface. The approach was a natural one for Newton to take given that his starting point was the *Principles of Philosophy,* in which a corpuscular ontology was assumed throughout. *De gravitatione* marks an abrupt change from this insofar as Newton treats fluids as continua. That assumption was made automatically and without comment in the hydrostatics of Stevin, Pascal and Boyle, as, indeed, it had been by Archimedes. As well as assuming continuity, the former three authors all accommodated into their theories the recognition that

[7] As Shapiro spells out in detail, Descartes' view that light is a pressing transmitted through a fluid medium (the second element or ether presumed to fill the space between Sun and earth) clashes with the rectilinear transmission of light once the isotropy of pressure is appreciated.

hydrostatic forces are isotropic. This was the key to the way in which they attempted to explain a range of hydrostatic phenomena including a number of otherwise puzzling ones. Continuity and isotropy are central to the hydrostatics of *De gravitatione* and alien to Descartes' *Principles of Philosophy.* I suggest that Newton's switch from considerations of interacting particles to those involving continuous fluids came about as a result of his becoming aware of the hydrostatic tradition. I further suggest that it is likely that he was alerted to that tradition by reading Boyle's *Hydrostatical Paradoxes*.

The reason I frame my ascription of an influence of the hydrostatic tradition on Newton, probably via Boyle, as a suggestion rather than assertion is that, as far as I am aware, there is no documentary evidence that Newton read Boyle's *Hydrostatical Paradoxes* or the work of Stevin and Pascal that is discussed and built on in that text. As Shapiro (1974, pp. 282–283) has stressed, 'there are sufficiently many similarities between Newton's and Stevin's approaches to suspect that Newton's hydrostatics derived at least in part from Stevin's. For his part, Pascal had identified continuity and fluidity as the key features of liquids, the latter serving to distinguish them from solids, echoing Newton's adoption of continuity and his struggle to formulate a definition of fluidity on which to base his hydrostatics. In Chap. 8 I argued that the theoretical device of conceiving of forces acting on either side of imaginary planes in the body of a fluid, freely adopted by Newton, appears for the first time in Boyle's *Hydrostatical Paradoxes,* a work which was a commentary on and extension of Pascal's hydrostatics. Whatever the extent of Newton's debt to the hydrostatic tradition, any seventeenth century figure seeking to adopt and exploit Newton's hydrostatics would have been able to appraise and augment it by incorporating the theoretical and experimental innovations of that tradition, so the relationships between the two are important irrespective of the extent to which Newton was indebted to the contributions of his predecessors.

9.4 Foundations of Hydrostatics

As I have already indicated in Sect. 9.2, in the opening paragraph of *De gravitatione* Newton distinguished two methods for treating the equilibrium of fluids. One was the way characteristic of a mathematical science and involved the derivation of propositions 'strictly and geometrically' from 'abstract principles sufficiently well known to the student'. This way, which typically abstracts from 'physical considerations', is contrasted with the way that involves deploying the mathematically derived propositions to explain natural phenomena thereby establishing its usefulness and 'perhaps' confirming its principles. The few pages of text that Newton eventually devoted to a mathematization of the equilibrium of fluids followed the first way. Newton made it clear that he was not averse to appeals to experiment involved in the second of the two methods but insisted that the two methods should not be confused.

In accordance with the dictates of the first method, Newton explained that 'the foundations from which this science [the equilibrium of fluids] may be demonstrated are either definitions of certain words; or axioms and postulates denied by none' (Hall and Hall 1962, p. 122). His subsequent endeavors to construct his hydrostatics show that formulating an adequate definition of 'fluid' was of crucial importance and far from straightforward, neither of which were conveyed by Newton's phrase 'the definitions of certain words'. What was needed was not some conventional definition of the word 'fluid' but a definition which served to capture the distinguishing features of fluids and from which, in conjunction with unproblematic postulates, the propositions of hydrostatics could be derived. We will be following Newton's attempts to frame a definition of 'fluid' that met these requirements below.[8] While Newton freely deployed the term 'fluid', replacing that term with 'liquid' would make more sense to a modern reader, since Newton's discussion is explicitly restricted to incompressible fluids from the outset.

If I am right to suppose that Newton read Boyle's *Hydrostatical Paradoxes* then he would have been struck by a key feature of the hydrostatic theories discussed there, especially those of Stevin, Pascal and of Boyle himself, a point that would have been reinforced had he proceeded to read the relevant writings of Stevin and Pascal directly. The authors in question automatically assumed, without comment, that fluids are continua. This marked a stark contrast with the treatment of fluids that Newton had encountered in the writings of Descartes, where they were considered as systems of interacting particles, each one pressing on its neighbors. Newton saw the benefits, and even the necessity, of treating fluids as continuous and explicitly addressed the issue in *De gravitatione*.

When, after his lengthy digression on space and motion, Newton returned to the task of developing a theory of the equilibrium of fluids, we find him struggling to reconcile an atomic or corpuscular picture of the world with the assumption that fluids are continuous. His conviction that the fundamental structure of the universe is atomic, openly discussed and advocated in his early notebook *Certain Philosophical Reflections* around 1665, stayed with him throughout his life. In the text on hydrostatics under discussion Newton referred to the 'least particles' constituting solids and liquids, arguing that 'on account of the innumerable circumstances affecting [their] figures, motions and contexture' it is not possible to construct a mathematical theory at that level. On the opening page of *De gravitatione* he had made the point that to develop a mathematical science it is necessary to abstract from physical considerations to some degree. The abstraction introduced by Newton involved the notions of an absolutely hard and an absolutely fluid body. An absolutely fluid body 'has no small portion or particle which is not likewise fluid' (Hall

[8] Newton's characterization of the mathematical structure of his theory, with its propositions derived from definitions and unproblematic postulates, bears a strong resemblance to the position advanced by Stevin on the status of his hydrostatics. This fact adds to the circumstantial case that Shapiro has made for the claim that Newton had read and was influenced by Stevin.

and Hall 1962, p. 151). That is, the extension of a fluid is a mathematical continuum.[9] An absolute hard body is also continuous, being 'a single undivided and uniform body'. While both a fluid and an absolutely hard body are continuous, they differ insofar as the parts of a solid, however small, are not able to move with respect to each other. When a hard body moves it moves as a whole, with all its parts suffering the same displacement in the same direction and in the same time. By contrast the parts of a fluid, however small, are free to move with respect to each other, so long as they do not move apart in a way that would bring about changes in volume.

> I define the parts [of a fluid] not as being in motion among themselves, but only as capable of motion, that is, as being everywhere so divided one from another that, although they may be supposed to be in contact and at rest with respect to one another, yet they do not cohere as though stuck together, but can be moved separately by any impressed force and can change the state of rest as easily as the state of motion if they move relatively (Hall and Hall 1962, p. 151).

This can be read as Newton's elaboration of the common sense appreciation that a solid retains its shape when free of strong distorting forces, whereas a portion of liquid readily adapts its shape to a containing vessel or an immersed solid.

The elaboration by Newton of his understanding of fluids, cited above, was better suited to his purpose than the explicit definition on which he was elaborating. Definition 18, as it appears in Newton's manuscript, reveals some indecision on Newton's part. He crossed out part of his initial formulation and then added an additional two sentences in a marginal note. His first attempt read 'A fluid body is one whose parts yield to an impressed pressure all whose parts are mobile among themselves' but he crossed out the last seven words referring to the mobility of the parts.[10] He also added a marginal note that read:

> Moreover, the pressures by which the fluid is driven in any direction whatsoever (whether these are exerted on the external surface alone, or on the internal parts by the action of gravity or any other cause), are said to be balanced when the fluid rests in equilibrium. This is asserted on the assumption that the pressure is exerted in some one direction and not towards all at once (Hall and Hall 1962, pp. 148–149).

I suggest that the deleted phrase is witness to the transition that Newton was making, from the corpuscular characterization of fluids he had inherited from Descartes to the continuum view that he was in the process of adopting and exploiting. Descartes had formulated his distinction between solids and fluids in Section 54 of Part II of *Principles of Philosophy*. He acknowledged that the senses reveal such a distinction, which we recognize to consist 'in the fact that the parts of fluid bodies easily move out of their places, and consequently do not resist the movement of our

[9] Newton pointed out that the assumption of continuity in play here is akin to that involved in the assumption that materials have a density. The mathematics presumes continuity, but 'in physics things seem otherwise' (p. 150) insofar as matter is made up of atoms.

[10] In their translation of Definition 18 Hall and Hall write 'overwhelming pressure' which I have replaced with 'impressed pressure'. (The Latin is '*praepollenti pressioni*'.) In the text that follows the definition Newton made it clear that adjacent parts of a liquid can be 'moved separately by any impressed force'. Even a solid will yield to a force that is sufficiently 'overwhelming'.

hands into those places, while, on the contrary, the parts of solid bodies adhere to one another in such a way that, without sufficient force to overcome their cohesion, they cannot be separated'. Descartes proceeded to give his account of what lies behind this common sense distinction, specifying that 'those bodies which are divided into very small parts which are agitated by a diversity of movements, are fluid; while those bodies whose particles are all contiguous and at rest, are solid'. Newton's initial formulation, prior to the deletion, echoes Descartes' definition. There are two reasons why he would have had second thoughts. Firstly, he was moving towards a continuum view according to which the 'parts' of a fluid are notional parts arising from the division of a continuum rather than discrete corpuscles. Secondly, the continuum perspective led him to the view that in a fluid in a state of equilibrium adjoining parts, however small, are at rest with respect to each other, and so not in relative motion. Parts of a fluid are freely capable of being set in motion by an impressed force, but are not in fact moving when the fluid is in equilibrium. The first sentence of Newton's marginal note can be read as stressing this point.[11]

The move from a corpuscular to a continuum account of the microstructure of liquids was not the only break with Descartes that can be traced in Newton's first venture into hydrostatics. A second break involved Newton's rejection of Descartes' categorical distinction between motion and rest that was noted in Chap. 5.[12] For Descartes, motion and rest formed a pair of logical contraries. A second pair of contraries was involved in Descartes distinction between directed motions, with motion in one direction being the logical contrary to motion in the opposite direction. This adoption by Descartes of the logic of contraries was deployed in his distinction between solids and fluids. For him the fact that parts of solids are at rest with respect to each other renders them difficult to separate in a way that is categorically distinct from the ease with which parts of fluids can be separated because those latter parts are already in motion. In his dynamics of interacting particles in the *Waste Book* Newton had already moved beyond Descartes insofar as he regarded

[11] The move by Newton from the corpuscular or atomic level to a macroscopic level that assumed continuity may well have been influenced by Boyle's 'New experiments physico-mechanical touching the spring of the air' of 1661. Newton's familiarity with that work is manifest by a number of references to it in his early notebooks. Near the beginning of *The Spring of the Air* Boyle referred to speculations concerning the properties of air that invoked the motion of its particles, citing Descartes in this connection. But he was at pains to emphasize that his notion of the 'spring of the air' that he invoked to explain the behavior of air, especially as revealed by experiments with his air pump, could be developed and supported in a way that was quite independent of assumptions about underlying corpuscles. 'Wherefore I shall decline meddling with a subject, which is much more hard to be explicated than necessary to be so, by him, whose business it is not, in this Letter, to assign the adequate cause of the spring of the air, but only to manifest, only that air has a spring, and to relate some of its effects'. Elsewhere, in Chalmers (2012), I have documented Boyle's articulation of a distinction between intermediate causes, such as weight and the spring of the air, and underlying atomic or corpuscular explanations of them, and the extent to which knowledge of the latter is not necessary for the pursuit of experimental science. Boyle's stand here matches the stand that Newton took on gravitational attraction which he invoked but could not explain at a deeper level. I have no direct evidence that Newton explicitly drew on this aspect of Boyle's work.

[12] As was noted in Chapter 5, Damerow et al. (2004, pp. 81–94) have given a detailed exposition of Descartes' adoption of a logic of opposites or contraries.

the force necessary to change velocity as proportional to the magnitude of the change independent of whether the change is from rest or from some degree of motion. From the point of view of Newton's dynamics, rest is not categorically distinct from motion but just a special case of it. This departure from Descartes is in evidence in Newton's characterization of solids, in a continuation of the passage from *De gravitatione,* cited above:

> Indeed, I suppose that the parts of hard bodies do not merely touch one another and remain at relative rest, but that they do besides so strongly and firmly cohere, and are so bound together, as it were by glue, that no one of them can be moved without all the rest being drawn along with it; or rather, that a hard body is not made up of conglomerate parts but is a single, undivided and uniform body which preserves its shape most resolutely, whereas a fluid body is uniformly divided at all points.

For Newton, rest is not sufficient to hold the parts of solids in place. Adjacent parts of fluids in equilibrium are at rest with respect to each other just as much as solids are. The difference between the two is that the adjacent parts of a fluid freely yield to forces whereas the adjacent parts of a solid do not, so that the solid 'preserves its shape most resolutely'. Newton's remark that adjacent parts of solids are held together 'as it were by glue' is presumably a counter to Descartes who explicitly denied that any glue was necessary. The distinction between fluids and solids can be carried over from a view that regards the parts as discrete particles to a continuum view where the parts can be as small as one wishes. Newton makes the transition to the continuum view in the closing clause of the passage quoted above.

I conclude this prelude to a look at Newton's technical development of his hydrostatics with some remarks on the second of the two sentences in the marginal note that Newton appended to his definition of fluid. Having asserted that the pressures driving a fluid 'in any direction whatsoever' are balanced when the fluid is in equilibrium, Newton declares that he makes this assertion 'on the assumption that the pressure is exerted in some one direction and not towards all at once'. It is by no means evident what Newton is getting at, but I can suggest what he is alluding to. If a force is communicated to a body of fluid in a closed vessel, by pressing or adding weights to a piston for example, then the liquid responds to the directed force by reaching a new equilibrium state in which the pressures within the fluid 'in any direction whatsoever' are balanced, though more intense than prior to the application of the force on the piston. A directed force of the kind referred to in Newton's second sentence is transformed by the fluid into isotropic intensity of force acting equally in all directions once equilibrium is attained. This circumstance was well appreciated and subject to experimental investigation in the tradition to which I am assuming Newton had access initially through Boyle, and it is a consequence of Newton's own theory. But since it is a consequence, it is out of place in a definition of a fluid, and sure enough, it is dropped from the definition appearing in the *Principia.*

9.5 Theory of Hydrostatics in *De gravitatione*

When Newton set out to formally deduce 'propositions on non-elastic fluids' in *De gravtatione* he began by considering a 'non-gravitating fluid' (Hall and Hall 1962, p. 152). He explicitly separated considerations of pressure from considerations of weight, thereby freeing himself from a focus on the latter that had impeded the development of hydrostatics from Archimedes to Descartes. His starting point was the most simple and symmetrical case of an incompressible, non-gravitating fluid in equilibrium, a sphere of fluid pressed equally and normally at all points of its outer surface.

Newton begins by offering a proof that a sphere of fluid compressed in such a way will be in equilibrium and motionless throughout. The fact that Newton felt the need to prove this is indicative of the extent to which, in his hydrostatics, he, like Stevin before him, aspired to the Euclidean ideal by striving to derive his theory from evident postulates 'sufficiently well known to the student'. Newton considered the segment CGEH of liquid bounded by two spherical shells and a cone with apex at the center, K, of the sphere (Fig. 9.1).

Newton argued that if the segment moved towards the center then all other similar segments situated at the same distance from the center, K, would also move towards the center because of the symmetry of the situation, contradicting the assumption that the fluid is incompressible. A similar argument was appealed to by

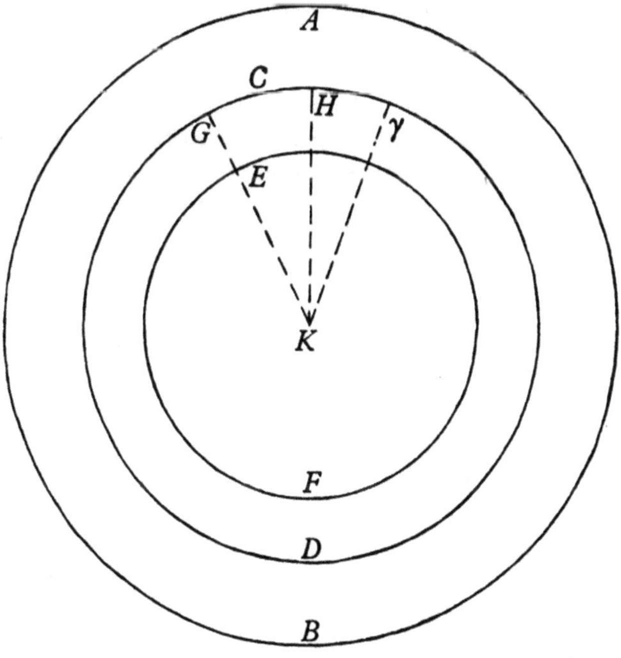

Fig. 9.1 Newton's sphere of liquid pressed equally around its surface (Hall and Hall 1962, p. 153)

Newton to establish that a segment like CGEH will not move radially outwards. His argument to the effect that the segment is unable to circulate, towards H, was less convincing. He claimed that if the segment moved towards H (that is, clockwise) then a neighboring segment, touching CGEH at H, would also move towards H (that is, anti-clockwise) which is ruled out by the incompressibility of the fluid. Newton has neglected the possibility that the fluid might rotate as a whole in a clockwise or anticlockwise direction. After all, the segment adjacent to CGEH, being similar to it, should be expected to execute a similar motion, that is, clockwise rather than anticlockwise. In the *Principia* Newton accordingly replaced the invalid argument with an argument that, in effect, rested on the assumption that there is no reason why the fluid should rotate in one direction rather than the reverse direction. Let us not pursue this matter further and concede to Newton the assumption that the sphere of liquid equally pressed on all sides will rest in equilibrium with all its parts motionless.

Newton proceeds to show that all the parts of the uniformly compressed liquid 'press each other equally or with equal intension' (Hall and Hall 1962, p. 152). Before I discuss his proof a small digression is necessary concerning Newton's terminology. Newton has adopted the traditional distinction between the 'intension' and the 'extension' of qualities. The intension of pressure corresponds to the force per unit area and the extension as the area over which it acts, the total force arising from the pressure being the product of its intension and its extension. The intension of pressure corresponds to what is called pressure in modern parlance and in the terminology that Newton had already adopted by the time he wrote the *Principia*. In the terminology of *De gravitatione*, Newton's objective is to show that the intensity of pressure is the same throughout the sphere of liquid in equilibrium and equal to the intensity of pressure applied at its surface. Let me now proceed to his proof which, as he presented it, is brief and unsatisfactory.

Newton considers the segment of water contained between a part, PQR, of the outer spherical surface, and a second spherical surface, PSQ, having the same radius as PRQ and so mirroring it. (See Fig. 9.2.) Newton's proof is as follows:

> For I have already shown that this part of the fluid remains in equilibrium, and so the effects of the pressures acting on both of its surfaces are equal, and hence the pressures are equal.
>
> And thus since spherical surfaces such as PSQ can be described anywhere in the fluid AB, and can touch any other given surface in any points whatever, it follows that the intension of the pressure of the parts along the surfaces, wherever placed, is as great as the pressure on the external surface of the fluid (Hall and Hall 1962, pp. 153–154).

Given that Newton has shown that the sphere of liquid is in equilibrium throughout, he can conclude that the force on surface PRQ is equal to the force on surface PSQ. But this is not sufficient to show that the intensity of pressure (that is force per unit area) on any location on surface PSQ is equal, and equal to the intensity of pressure at all locations on surface PRQ. As it stands, Newton's proof would apply as much to a solid sphere as to a liquid one. Each part of a solid sphere, which is

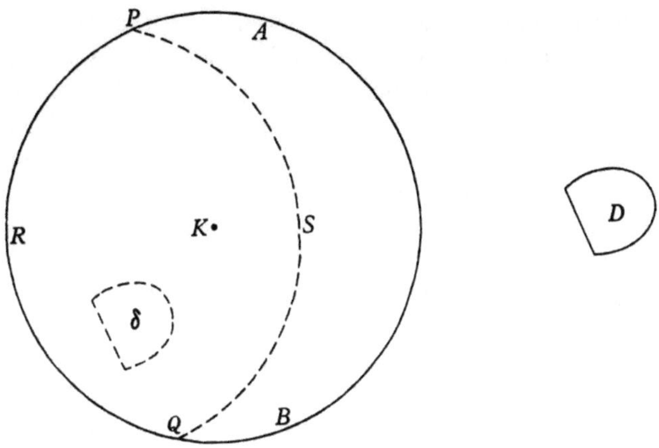

Fig. 9.2 Diagram for Newton's analysis of pressure in *De gravitatione* (Hall and Hall 1962 p. 154)

pressed equally on every point of its outer surface, will be in equilibrium for the same reasons as the parts of the fluid are, as Newton's opening argument showed. But on a surface such as PSQ within a *solid* sphere the forces will not be evenly distributed over the surface. The forces on the outer surface, which are directed towards the center of the sphere, will urge all parts of the solid in that direction. Consequently the forces on the surface of PSQ other than at point S will not act normal to the surface. Each part of the solid must react to these forces by offering resistance to them.

The fact that Newton did not invoke the difference between solids and liquids in his proof suggests how it can be reconstructed so as to change it into a valid one. Newton specified a fluid body as one 'whose parts yield to an impressed force' and which parts 'can be moved separately by any impressed force'. This can be read as implying that the forces on surfaces PRQ and PSQ must be symmetrical, for any asymmetry will constitute forces to which the liquid bounded by the surfaces must yield. On this reading, the intensity of pressure at each location on PSQ must be equal to the intensity of pressure at the corresponding point on PRQ and the force must be normal to the surface at that location since the force at the corresponding point on PRQ is normal.[13]

Newton pointed to the generality of the proof he had given with respect to the surface PSQ by noting that 'spherical surfaces such as PSQ can be described anywhere in the fluid, and can touch any other given surfaces at any points whatever' (Hall and Hall 1962, pp. 153–4). This point can be spelt out more explicitly than Newton did. Given an arbitrary plane through an arbitrary point within the sphere of liquid, ABR, a sphere of radius equal to that of ABR can be constructed which

[13] Shapiro (1974, p. 281) has noted that Newton's brief proof lacks validity and suggests that it can be rendered valid by adding the assumption that the forces on PSQ act normal to the surface. Since, as Shapiro acknowledges, Newton nowhere states such an assumption I have preferred to render his argument valid by drawing on the definition of a liquid that Newton did state.

passes through that point and has the plane as tangent. The constructed sphere will cut the original sphere ABR to generate a volume bounded by two spherical segments of equal radius, similar to those represented in two dimensions by the arcs PSQ and PRQ in Newton's diagram. Newton's proof (or my charitable interpretation of it) can now proceed as before to yield the general conclusion: on any surface through any point in the body of the sphere of liquid there is an 'intensity of pressure' equal to the intensity of pressure on the outer surface of the sphere. This has the consequence that at any point in the body of fluid there is 'an equal degree of pressure in all directions' as Newton stressed in the penultimate sentence of *De gravitatione*.

An implication of the discussion of hydrostatic equilibrium in *De gravitatione* was highlighted by Newton. He noted that the sphere of fluid AB figuring in his proofs, held in place by a spherical solid surface, could just as well be held in place by the surface of a further expanse of fluid surrounding and confining sphere AB. 'For the method by which sphere AB is compressed is of no significance, so long as its compression is supposed to be equal everywhere' (Hall and Hall 1962, p. 154). The replacement of solid confining surfaces by liquid ones and vice versa was deployed to greater effect in his treatment of hydrostatics in the *Principia*.[14]

9.6 Hydrostatics in the *Principia*

The version of hydrostatics appearing in the *Principia* is presented without fanfare and is condensed into two Propositions, 19 and 20, occupying just six pages of Cohen's translation of Section 5 of Book II. The formal, geometrical treatment presupposes that fluids are continuous, although that assumption is not stated explicitly. Newton freely uses, but does not explicitly define, the term 'pressure' and it is clear from the context that he is using it to denote what he had referred to in *De*

[14] Shapiro (1974, p. 280) refers to this assumption as 'Newton's principle of liquefaction' and likens it to the 'principle of solidification' that some have read into Stevin's *Elements of Hydrostatics'*. I avoid talk of such principles because I believe it is misleading from a historical and logical point of view. Neither principle is explicitly formulated, either by Stevin or Newton. The principles of solidification is presumed by Stevin in the sense that adding it to his postulates renders some of his proofs valid which are otherwise invalid. But that move would have been problematic for Stevin insofar as he required his postulates to be sufficiently unproblematic to be granted at the outset of his deliberations. A principle of solidification could no more be taken as an evident starting point for Stevin than a principle of liquefaction could be taken as a starting point for Newton. The replacement of solid confining surfaces by liquid ones was sanctioned by Newton as a consequence of the fact that equilibrium requires forces to be balanced, together with some features of solids and liquids that did form the basis of his hydrostatics. It is therefore misleading to talk of that assumption as a 'principle'. The first explicit formulation of a general principle in this context was the principle of solidification formulated by the French mathematician Alexis Clairaut in 1743, long after Stevin and Newton had made their respective contributions to hydrostatics. On this latter point see Truesdell (1954, pp. xi and xix–xx). An engineer's perspective on the history of this issue can be found in Casey (1992). I owe this latter reference to Alan Shapiro.

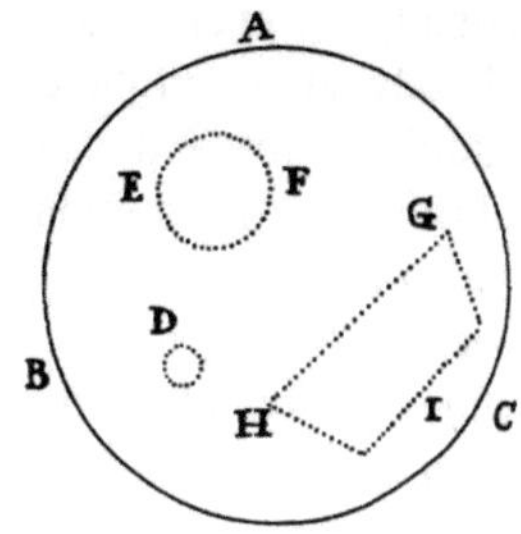

Fig. 9.3 Diagram for Newton's treatment of pressure in the *Principia* (Cajori 1962, p. 290)

gravitatione as the 'intensity of pressure'. Pressure in a particular direction is now, in the *Principia* version, force per unit area in that direction, and when Newton talks of bodies 'equally pressed on every side' it is clear that he means bodies subject to the same pressure (force per unit area) on every side. As was the case in *De gravitatione*, Newton clearly separated considerations concerning pressure (Proposition 19) from those concerning weight (Proposition 20). Discussion of the former Proposition is confined to technical detail unaccompanied by meta-discussion of the kind appearing in *De gravitatione*, It is improved by an explicit appeal to the definition of a liquid that was lacking in the earlier work and by explicit appeal to Newton's third law expressing the equality of action and reaction. The extension to gravitating fluids in Proposition 20 goes beyond the considerations of *De gravitatione* and includes Newton's treatment of the 'hydrostatic paradoxes' that had posed the challenges to which the hydrostatics of Pascal and Boyle had been a response.

According to Proposition 19 all parts of an incompressible fluid in equilibrium, free of gravity and all centripetal forces, are 'equally pressed on every side', that is, subject to the same force per unit area on every side. The key premise for the demonstration of the proposition is the definition of a fluid:

> A fluid is any body whose parts yield to any force applied to it and yielding, are moved easily with respect to one another (Newton 1999, p. 687).

As in *De gravitatione,* Newton begins by considering a fluid ABC (Fig. 9.3) contained in a spherical vessel and equally pressed at all points of its surface, and once again concludes that any portion of the sphere of fluid will remain at rest. Motion towards or away from the center by any part will entail condensation of the fluid, contrary to the assumption of incompressibility, whilst a circulating motion can be ruled out because there is no reason why the circulation should take place in one direction rather than the other.

Immobility having being established, in a way that would apply to the parts of a solid body as much as a fluid one, Newton proceeds to prove that all the parts of the sphere of *fluid* compressed uniformly are themselves pressed equally in all directions. He considers a small spherical part or sub-sphere of the fluid, EF, recognizing that he has already established that such a sub-sphere will be motionless if the fluid is in equilibrium. Newton entertains the assumption that the sphere EF is not pressed equally on every side and supposes that an additional force be introduced to rectify the imbalance. Once the imbalance is rectified the sub-sphere is a sphere pressed equally on every side, and so its parts must remain motionless, as Newton has already shown in the case of the large sphere. But, prior to the application of the additional force, the sub-sphere and its parts were already motionless and in equilibrium, so the additional force must cause some relative motion of the parts of the sub-sphere 'by the definition of a fluid' (Newton 1999, p. 687). The fluid must move because the forces on it are not balanced and because, being a fluid, it is unable to withstand unbalanced forces. The assumption that the sub-sphere is not pressed equally on every side has led to the contradictory conclusion that its parts must both move and not move.

Once again, Newton's argument concerning fluids can be clarified by pinpointing the way in which the argument does not carry through for solids. In the latter case, applying the hypothetical equalizing forces involved in Newton's argument to a *solid* sub-sphere will not induce motions in the way that they would do in the case of fluids. Rather, the solid, whose adjacent parts are not free to move with respect to each other, will simply resist the applied forces, with unevenly distributed internal tensions resulting. I should point out that in the *Principia* version of his hydrostatics Newton did not explicitly make the point that he had made in *De gravitatione*, namely, that adjacent parts of a solid 'are so bound together, as it were by glue, that no one of them can be moved without all the rest being drawn along with it', but his case implies such an assumption.

Let us proceed to follow Newton's development of the point, now established, that on any sub-sphere of a larger, uniformly compressed, fluid sphere the pressures are uniform and normal to its surface. This result now opens the way for Newton to show that on any plane, whatever its orientation, through any point within the sphere of fluid, the fluid presses equally on either side of it with a pressure that is equal to that on the outer surface of the sphere. Given any point and any plane containing it, consider a small sphere that touches the plane at that point and consider a second sphere of such a radius that it touches the small sphere at the point in question and also the side of the containing vessel. Newton has demonstrated that each of the spheres, like the large sphere enveloping them, must be equally pressed on every side. He also concludes from the fact that the entire fluid is in equilibrium that the forces of the sub-spheres on each other must be equal and opposite as must the force between the container and the sphere that touches it. Newton explicitly invokes his third law of motion, that action and reaction are equal and opposite, to justify this latter move. The conclusion is that the sub-spheres press on each other in a direction perpendicular to their common tangent with a pressure that is equal to that with which the sphere touching the containing surface presses on and is pressed by it. In short, the pressure is uniform throughout the fluid. What is more, the pressing acts in all directions simultaneously, because spheres can be drawn tangential to any plane through any given point and the above argument repeated for each of them.

The remaining content of Newton's hydrostatics for non-gravitating fluids is shown by him to follow from the pressure being uniform throughout the body of a uniformly compressed sphere and that equilibrium results from balancing forces, the latter being a consequence of his third law of motion. Newton was able to exploit the point, already made in *De gravitatione*, that, as far as the equilibrium of a body of fluid is concerned, it makes no difference to that equilibrium whether the fluid is held in place by a solid surface or by a further expanse of fluid. The pressure on the surface of the unevenly shaped body of fluid GHI depicted in Newton's diagram (Fig. 9.3) is equal to that on the outer surface of the sphere and that same pressure is evenly distributed throughout GHI. Equilibrium is sustained because the force exerted *on* GHI by the pressure of the surrounding fluid is countered by the force exerted *by* GHI on it. As far as the equilibrium of GHI is concerned, in makes no difference whether it is held in place by a solid surface or by a liquid one. In each case, the pressure exerted on the surface is countered by an equal and opposite pressure. So the pressure in a fluid contained in a vessel of any shape, not just spherical, is equally distributed and equal in magnitude throughout to the force per unit area on the surface. Conversely, the pressure on the surface of a solid immersed in a fluid will be equal to the pressure throughout that fluid.

In the penultimate paragraph of the discussion of pressure in non-gravitating fluids, Newton summarized his conclusions in a way that stressed their dependence on the nature of fluidity and the equality of action and reaction.

> And thus in a rigid vessel, a fluid will not sustain a pressure that is greater on one side than on another, but will yield to it, and will do so in an instant of time, since the rigid side of the vessel does not follow the yielding liquid. And by yielding, it will press the opposite side, and thus the pressure will tend on all sides to equality. And since, as soon as the fluid endeavors to recede from the part that is pressed more, it is hindered by the resistance of the vessel on the opposite side, the pressure will be reduced on all sides to equality in an instant of time and without local motion; and thereupon the parts of the fluid—will press one another equally and will be at rest with respect to one another (Newton 1999, p. 688).

Here Newton forcefully expresses the, somewhat counter-intuitive, way in which fluids, by virtue of being such, convert directed forces impressed upon them into isotropic ones. He stresses the way in which this is brought about as a consequence of the forces at work due to the fact that the fluid is free to adapt to any impressed force and to the reaction it induces from any solid surface that impedes it. In effect, Newton gives a mechanistic account of the transmission of forces through liquids that I have argued to be lacking to some degree in earlier theories of hydrostatics. Newton has explained how it is that liquids can push around corners.

9.7 Extending Hydrostatics to Include Weight

Armed with the account of the isotropy of pressure that he has established by proving and elaborating on his Proposition 19, Newton proceeded to prove Proposition 20 which concerns gravitating fluids. Newton considered a homogeneous fluid distributed symmetrically about a solid spherical surface with each part of said fluid gravitating towards the center with a force that is the same at equal distances from that center. He showed that the total force on the solid surface is equal to the weight of a cylinder of fluid with a cross section equal to the area of the surface, and a height equal to that of the upper surface of the fluid above the solid surface on which it presses.

Newton's proof of Proposition 20 utilized what he had established in Proposition 19, that any force applied to a fluid is communicated isotropically through the body of the fluid as a uniform force per unit area, that is, as a uniform pressure. Newton imagined the gravitating fluid to be divided into concentric spherical shells. The outer shell, by weighing on the adjacent shell would cause a uniform pressure on the outer surface of that second shell which will be transmitted, as the same force per unit area, to its inner surface. The upper surface of the third shell will now be pressed not only by this pressure, but also by a pressure due to the weight of the second shell, and so on. By the time we reach the solid spherical surface supporting the fluid, the pressure will have been augmented by the pressure (weight per unit area) of all the layers above it. 'Provided the number of shells is increased and their thickness decreased indefinitely' (Newton 1999, p.689) the force per unit area contributed by each shell will be the weight of a cylinder of fluid of unit cross sectional area and a height equal to the infinitesimal thickness of the shell. Consequently, the force per unit area on the solid surface will be equal to the weight of a cylinder of fluid of unit cross section and a height equal to the distance separating the upper surface of the fluid from the solid surface supporting it. The total force on the latter surface is obtained simply by multiplying the force per unit area by the area of that surface. This yields the result that Newton set out to prove.

Since Newton's proof relies only on the isotropic transmission of pressure and the spherical symmetry of the gravitating force, it is valid whatever the function is that relates gravitational force to distance from the center. Newton's proposition can certainly be applied to cases where the gravitational force is constant, as can be assumed as an approximation for volumes of fluid encountered near the Earth's surface, and for cases where gravitational attraction varies inversely with height, as in astronomical settings.

Having established his proposition Newton could employ it to explain the phenomenon of floating, the loss of weight experienced by immersed solids, and a range of' paradoxical' hydrostatic phenomena to which seventeenth century hydrostatics had been designed to respond. For instance, as we saw in Chap. 3, already in the late sixteenth century Stevin's hydrostatics included the result that the force exerted on the end surfaces, EF, of each of the tubes of water depicted in Fig. 3.1, p. 31, is the same provided the depth of those surfaces beneath the upper surface of the water, MI, and the area of the surface at EF is the same in each case. This remains the case whatever the contours and varying cross-sections of the tubes and hence whatever the total weight of the water in them might be. As Newton indicated, the method of proof he employed to establish Proposition 20 can be adapted to such cases. The water in each tube can be divided into thin horizontal layers. The way in which weight per unit area is accumulated in the progression from the upper layer at MI to the one adjacent to EF can be followed through in a way analogous to the way in which Newton had treated consecutive layers in the spherically symmetric case. In cases where the tube extends below EF before rising again to meet it, the weight of consecutive ascending layers will simply need to be subtracted from, rather than added to, force arising from the pressure. Newton's line of argument leads to the result that the force on EF is equal to the weight of a cylinder pf water of cross-section equal to the area at EF and a length equal to the vertical height of MI above EF. Newton derived Stevin's law.

9.8 Newton's Hydrostatics in Perspective

A cursory perusal of the texts on hydrostatics written by Stevin, Pascal, Boyle and Newton indicates a similarity between the first and last of these that sets them apart from the middle two. Geometrical proofs aided by diagrams figure crucially in both Stevin's *Elements of hydrostatics* and the two texts on hydrostatics by Newton analyzed in this Chapter. This signals the way in which the two authors aimed to provide formal accounts of hydrostatics the contents of which took the form of theorems derived as deductive consequences of postulates cast in a geometrical form. Newton succeeded where Stevin had failed insofar as the former's postulates, unlike Stevin's, were in the main capable of yielding the content of hydrostatics as valid consequences. Newton was able to take advantage of advances made by Pascal and Boyle and incorporate their physical insights into his formal geometrical theory. In short, Newton put hydrostatics on a secure footing by identifying the postulates from which it could be derived, the most important of which involved an adequate characterization of the difference between fluids and solids.

The relevant contributions to hydrostatics by Pascal and Boyle on which Newton was free to draw can be summarized as follows: Pascal, in his treatise *On the Equilibrium of Liquids*, gave clear expression to the way in which directed forces applied to a body of liquid at one location on its bounding surface result in the same force per unit area on other bounding surfaces however remote and whatever their orientation. He illustrated his claim, sometimes referred to as 'Pascal's principle', by reference to what has become known as the hydraulic press. Pascal (1937, p. 7) talked of water pressing equally on the two pistons in the hydraulic press, leading to forces proportional to their respective areas, but he did not use the term 'pressure' (*la pression*) and did not discuss forces acting in the body of the water in the press. In his commentary on and extension of Pascal's theory. Boyle, in *Hydrostatical Paradoxes*, did make explicit the notion of pressure. In a subsequent text, *A Hydrostatical Discourse*, published in 1672, he dramatized the distinction between pressure and weight as the causes of hydrostatic phenomena by noting how a solid can be made to float in a miniscule amount of water, a circumstance that demands that the cause of floating be attributed to pressure as opposed to weight.[15] Boyle considered the way in which pressure acts within the body of a liquid in a more explicit way than Pascal had. In doing so he introduced the theoretical device of considering the forces acting on either side of imaginary planes within the body of a liquid. Boyle made no attempt to formulate his hydrostatics mathematically. When

[15] See Boyle (1999, pp. 161–164).

Pascal (1937, p. 8) did depart from his largely physical exposition to make a mathematical point he downplayed its significance, describing it as an issue that could be understood by geometers but which could be disregarded by others.

By moving beyond the physical insights of Pascal and Boyle, Newton was able to fulfill Stevin's aspirations and cast hydrostatics into the form of a geometrical system of theorems derived from postulates sufficient for the purpose. The key postulate was that which identified the distinction between a fluid and a solid. Solids resist distorting forces whereas fluids give way to them. Making free use of the device of considering the forces on hypothetical surfaces within the body of a liquid which Boyle had pioneered, Newton fashioned the technical notion of pressure to develop the consequences of this distinction. In doing so he made explicit for the first time that, at any location within the body of a fluid in equilibrium, pressure acts equally in all directions. Newton established his conclusions about pressure prior to investigating the consequence of the fact that fluids have weight. Once he did include considerations of the latter kind he was able to show that a body of liquid subject to forces directed towards a common center, of which weight is a special case, is subject to no movement as a consequence of those forces (Newton 1999, p. 690). A body of liquid, whether constrained in a container or free to spread like the oceans, and acted on only by a central force such as gravity, will rest in equilibrium with all parts of its upper surface equidistant from the center.

Newton ended his discussion of hydrostatics in the *Principia* with nine corollaries that summarized its empirical content. He explained the apparent loss of weight of immersed solids, floating, the hydrostatic paradoxes and the fact that insects under water are not distorted by pressure. That is, Newton departed from what he had called the first way in *De gravitatione*, the mathematical proofs of hydrostatic propositions, to the second way, the illustration of hydrostatics by way of its application to a range of phenomena. Many of those phenomena, such as floating, had been familiar for many centuries. Others had increasingly become common knowledge since the deliberations of Stevin a century earlier and had been highlighted by a multitude of easy-to-perform experiments by Boyle, many of which were described in his *Hydrostatic Paradoxes* of 1666. Newton's achievement was not to add content to hydrostatics but to identify the assumptions on which it depended and to demonstrate its content as mathematical consequences of those assumptions.

I end by noting two respects in which Newton can be said to have left his task incomplete. They relate to the definition of a fluid and the extension of hydrostatics to include air as well as liquids.

We have seen above that in *De gravitatione* there is evidence that Newton prevaricated over his definition of a fluid. The part of his initial effort that survived in the final version reads 'A fluid body is one whose parts yield to an impressed pressure'. He subsequently spelt this out by noting that adjacent parts of a fluid 'can be moved separately by any impressed force'. Yet this is still not enough. Central to Newton's own discussion, both in *De gravitatione* and in the *Principia*, is an example of forces on a fluid that do not cause their adjacent parts to move. Those forces are those that are directed so as to uniformly compress the fluid. It is distorting forces rather than compressive forces that fluids are unable to oppose. This is implicit in Newton's deliberations and is a consequence of the fact that the force due to pressure on any plane within the body of a fluid in equilibrium is normal to that plane, a consequence that followed from Newton's discussion but was not highlighted by him. It was left to Leonhardi Euler to do that.[16] The isotropy that is characteristic of pressure in fluids follows from this.

Newton's discussion of pressure in *De gravitatione* and his derivation of the isotropy of pressure in the *Principia* work on the assumption that fluids are incompressible. Newton did not show how his treatment could be extended to compressible fluids, that is, to air. When, in the section on Proposition 20 of the *Principia,* he extended his discussion to include pressures arising from gravity, Newton did not restrict his claims to incompressible fluids, although he did not explicitly acknowledge the change. In his proof of Proposition 20 Newton in fact assumed Proposition 19. But he could not legitimately do so in the case of compressible fluids because he had not proved the latter proposition in the more general case. For this reason, Newton's hydrostatics fell short of being a unified theory of fluids in equilibrium governing liquids and air alike. Once again, it was Euler who completed the job.[17]

9.9 The Dating of *De gravitatione* Revisited

The innovations in mechanics made by Newton began their emergence as the result of his detailed engagement with Descartes' *Principles of Philosophy*. There are two features of the hydrostatics of *De gravitatione* that could not have arisen by such a route, the intrinsic connection between the transmission of hydrostatic forces and the strict continuity of liquids, on the one hand, and the separation of pressure and weight, on the other. The germs of both of those notions were present in the hydrostatic tradition that must have become known to Newton via Boyle's *Hydrostatic Paradoxes* of 1666, if it had not already done so by some other route.

The mechanics that had emerged in Newton's thought towards the end of the 1660's was what Ruffner (p. 249) has aptly termed 'billiard-ball physics' insofar as it treated of particles or corpuscles that maintained their linear motion or rest unless pushed by or colliding with other particles or corpuscles. It shared this feature with Descartes' system notwithstanding the crucial difference that Newton allowed for particles moving in a void that was antithetical to Descartes' plenum. To the extent that the two versions of billiard-ball physics could be identified with metaphysical

[16] Euler's 'General principles of the state of equilibrium of liquids' was published in 1757 (Euler 1954, pp. 2–53). A partial English translation of and commentary on it is included in Truesdell (1954, pp. lxxv–lxxxiii). Euler identified the 'essential property of fluids' by specifying that 'a fluid mass cannot be in equilibrium unless it is subject at all points of is surface to forces equal and perpendicular to the surface' (Truesdell 1954, p. lxxvi).

[17] Euler (1954) developed a general theory of hydrostatics applying to both compressible and incompressible fluids. His derivations were facilitated by his ability to argue algebraically using partial differential equations rather than being restricted to geometrical arguments aided by diagrams, as Newton had been.

world views they stood opposed to continuity. Nevertheless, Newton came to recognise that the way in which liquids were presumed to be continuous in the hydrostatic tradition was in fact essential for a science of hydrostatics that embraced the isotropy characteristic of pressure. Particles of solid, however small, tend to move as a whole in the direction of an impressed force because their parts cohere. Particles of liquid, *however small,* do not do so because their parts do not cohere insofar as they offer no resistance to distorting forces. Liquid are liquids all the way down. In light of this distinction it can be appreciated that hydrostatics cannot be reduced to systems of corpuscles like billiard balls pressing against each other, as Descartes had attempted to do. Newton explicitly discussed this issue in *De gravitatione* as we have seen, stressing that a fluid 'has no small portion or particle that is not likewise fluid'.

As we saw in Chap. 5, Descartes attributed hydrostatic forces to corpuscles weighing on corpuscles. He likened this to the transmission of light via the pressing of ether particles on their neighbours due to the centrifugal forces generated by the vortex motions in which they participated. The hydrostatics of *De gravitatione* treats of pressure in a non-gravitating liquid. Such a move would had been unmotivated and out of place in Descartes' *Principles of Philosophy.*

The considerations of the previous two paragraphs make it imperative to understand the hydrostatics of *De gravitatione* as the fruition of the hydrostatic tradition stemming from Stevin to Boyle rather than as emerging from an engagement with the details of Descartes' *Principles of Philosophy*, as much of Newton's mechanics undoubtedly did. That is the case however and whenever Newton became aware of the hydrostatic tradition. Given Newton's close association with Boyle, it is most improbable that he was unaware of and had not read Boyle's *Hydrostatical Paradoxes* of 1666. We saw in Chap. 6 how Pascal made moves towards separating hydrostatic forces for the weight of the liquid through which they are transmitted and, in Chap. 8 how Boyle stressed that an adequate understanding of floating requires appeal to pressure in addition to and as distinct from weight. In the hydrostatics of *De gravitatione* we find Newton freely exploiting, automatically and without comment, the device of considering forces acting on either side of planes within the body of liquids, a move that Boyle had introduce in *Hydrostatical Paradoxes*, claiming priority for it. It would seem that, having read Boyle's *Hydrostatical Paradoxes* and perhaps also the works of Stevin and Pascal referred to in it, Newton grasped the essential role that the continuity of liquids plays in the isotropic transmission of forces through them and developed a mathematized version of hydrostatics that gave a central role to that feature. If this assumption is correct, then one can expect Newton to have composed the hydrostatics in *De gravitatione* within a small number of years after 1666, that is, by 1668 or shortly thereafter.

Much of the debate concerning the date of composition in the literature draws on the long digression about space and body that interrupts the hydrostatics. However, a focus on the latter helps to strengthen the view that Newton's discussion in the *De gravitatione* bears the marks of his detailed engagement with Descartes *Principles of Philosophy* that was characteristic of his work in the 1660s. For instance, his insistence that the parts of solids cohere because they are held together 'as if by glue' is clearly an allusion to Descartes' denial that any glue is necessary. For Descartes, solidity arises simply because neighbouring parts of solids are at rest with respect to each other, a position that was tenable only in the light of Descartes' insistence that motion and rest were logical contraries. While Newton may have been compelled to take Descartes' position seriously and distance himself from it in the late 1660s, it is hard to believe that he still felt that compulsion in the years that he was composing the *Principia*. By this time Newton's views on force and inertia had developed to a stage where rest could be understood as a special case of motion rather than a logical contrary of it. Newton's talk of density and pressure employing the scholastic distinction between the intension and extension was another practice that was outmoded by the 1680s. In the version of hydrostatics appearing in the *Principia* Newton simply takes the coherence of solids for granted, without comment, and talks of pressure of liquids rather than intensity of pressure as in *De gravitatione*.

We have noted that Newton opened his discussion of *De gravitatione* by distinguishing between the mathematical and experimental methods, declaring that he will utilise the former in his exposition of hydrostatics. Dobbs (1991, p. 141) has likened this to the distinction Newton makes in the Preface to the first edition of the *Principia* and includes it in her list of themes common to that work and *De gravitatione*.[18] It is this commonality that is at the basis of her claim that the latter work was composed around the same time as Newton was writing the *Principia*. However, the distinctions between the methods as they are formulated in the two locations are not the same. In the Preface in question, Newton distinguished between mathematically formulated natural philosophy and practical mechanics as involved in the practical arts, which he construes as too imprecise to be subject to mathematical treatment. He notes that the geometry of the ancients presupposes exact straight lines and circles abstracted from but not the same as the lines and curves implicated in the practical arts. Newton indicates that in the *Principia* he will construct a 'rational mechanics' that will extend mathematization of geometry by the ancients to the mathematization of forces and the motions they give rise to. The result will be a mathematical natural philosophy as opposed to manual arts.

> But since we are concerned with natural philosophy rather than manual arts, and are writing about natural rather than manual powers, we concentrate on aspects of gravity, levity, elastic forces, resistance of fluids, and forces of this sort, whether attractive or repulsive. And therefore our present work sets forth mathematical principles of natural philosophy (Newton 1999, p. 382).

I take the 'therefore' to imply that mathematization is possible in Newton's rational mechanics because it is concerned with the motions resulting from naturally occurring rather than manual 'powers'. This raises the 'basic problem' of how mathematized versions of the forces acting in nature are to be acquired. Newton is clear that these must be discovered 'from the phenomena'. The way in which this can be done is best exemplified by the case made for the law of universal gravitation

[18] A translation of Newton's Preface is in Newton (1999, pp. 381–383).

in Book 3 of the *Principia.* The argument is subtle and complex, but it is empirical insofar as it draws on observational and experimental evidence. The specification of universal gravitation cannot be abstracted from common sense practices as readily as the concept of straight line figuring in geometry can.

Once it is interpreted in this way, then Newton's distinction between rational mechanics and manual arts does not map onto the distinction between the mathematical and the experimental method as specified at the beginning of *De gravitatione*. There it is made clear that a mathematical science abstracts from physical considerations and involves theorems strictly derived from postulates 'well known to the student' and 'denied by none'. This mathematical method is contrasted with 'a freer method of discussion' which involves illustrating the propositions so proved 'abundantly from experiment'. The mathematical way that Newton intends to follow in his hydrostatics involves establishing it by abstracting from physical situations and deriving theorems from postulates already known. It stands opposed to a mode of arguing that appeals to observation or experiment in a way that the rational mechanics of the *Principia*, especially its dealings with gravity in Book 3, does not.

Newton's distinction between the two methods in *De gravitatione* makes immediate sense once we turn our attention to hydrostatics, which was, after all, the topic that he set out to address. The mathematical method can be taken as applying to the approach adopted by Stevin in his hydrostatics and contrasted with the experimental method adopted by Pascal and Boyle. Newton would have been alerted to the contrast between the style of reasoning of Stevin as opposed to that of Boyle and Pascal on reading Boyle's *Hydrostatical Paradoxes* and the works by Pascal and Stevin discussed there. Having noted the contrast he sided with Stevin. Where Stevin had started with postulates that were abstractions and idealizations sufficiently unproblematic to be granted, Newton began with propositions of a similar kind denied by none and known to the student. Both developed their theories by rigorously deriving propositions from their postulates and used geometrical diagrams to aid them. Both distinguished between the formal part of their reasoning and the practical application of its fruits. Once Newton had seen the way to derive the content of his hydrostatics from a concise definition of the liquid state he could discount the need for empirical support for it. This was a strategy that would prove insufficient when it came to the specification of gravitational attraction and other force laws. For this reason it makes more sense to date the composition of *De gravitatione* close to his becoming aware of the hydrostatic tradition and relatively remote from the trickier problems he confronted in the *Principia.*

The case for the early dating of *De gravitatione* can be further strengthened by drawing on the relation between Newton's hydrostatics and his theory of light as characterised by Shapiro (1974).[19] As Shapiro notes, in his notebook *Certain Philosophical Questions,* composed around 1665, Newton shows a detailed familiarity with Descartes' theory of light. One of the grounds on which he opposed it was the fact that, if light is transmitted by way of corpuscle pushing against

[19] Shapiro assumed a publication date of around 1668 for *De Gravitations*, a position that was totally reasonable in 1974, a decade before the case for a later date had begun to gain credence.

corpuscle, as Descartes maintained, then light will not necessarily move in straight lines. This is because the direction in which an impelled corpuscle tends to move will not be the same as the direction in which it is pushed by a neighbouring particle, accept in the special case that the corpuscles are aligned in the direction of that push. At that stage, it is apparent that Newton believed the departures from linearity implied by Descartes' theory would be small, but nevertheless sufficient to clash with the observed sharpness of the edges of shadows, eclipses, and the blocking of a beam of light by a small object. On Descartes theory 'a little body interspersed would not hinder us from seeing, pression could not render shapes so distinct, ye sun could not be quite eclipsed' (Westfall 1962, p. 174). Again as Shapiro has noted, by 1672 Newton had become aware that departures for linearity of light rays to be expected if the transmission of light is akin to a pressure will be drastic, not slight. There will be a 'drastic and very extravagant spreading' of light rays on passing through an aperture, for example.[20] It is highly probable that the change in Newton's stance was brought about by his becoming acquainted with the hydrostatic tradition and the concept of pressure with which it worked. What is more, Newton now had a motive for clarifying and improving that account of hydrostatics insofar as it aided his case against theories of light that likened it to a pressure and favoured his own particle theory.

My main aim in this Chapter has been to improve our understanding of Newton's hydrostatics by portraying it as a major step forward in the hydrostatics tradition from Stevin to Boyle. That feature of my case is independent of the precise dating of the composition of *De gravitatione*. Nevertheless, I also suggest that the assumption that it was written in 1668 or shortly thereafter makes more historical sense of the development of Newton's thought in mechanics, optics and hydrostatics than the assumption that it was written in the mid-1680s, at a time when Newton was composing the *Principia*. The hydrostatics of Stevin, Pascal and Boyle, like that of Archimedes before them, was engaged with the explanation of hydrostatic phenomena in a way that was not tied to a metaphysical world view and treated liquids as continuous in a common sense. On becoming familiar with that tradition Newton grasped the fact that the different behaviour characteristic of solids and liquids requires that liquids be treated as continua in a strict sense that went beyond the common sense. That move was at odds with the atomistic metaphysics that Newton favoured throughout his adult life and could not have been motivated by considerations at that level. Nor could the notion of continuity required by Newton's definition of a liquid have arisen from mechanical considerations of the kind central to the *Principia*. Rather, it arose as the result of considerations within and peculiar to the hydrostatic tradition, a tradition that Newton was probably introduced to via Boyle's *Hydrostatical Paradoxes* of 1666. Once the lengthy metaphysical digression is removed from *De gravitatione* what remains can be seen as Newton's formal mathematization of hydrostatics, made possible by his recognition that the essential property of liquids is that they cannot resist deforming forces, and that they possess that property 'all the way down'.

---

[20] Letter to Oldenburg in Turnbull (1959, p. 175).

References

Boyle, R. 1999. In *The works of Robert Boyle. 14 volumes*, ed. M. Hunter and E. Davis. London: Pickering and Chatto.

Cajori, F. 1962. *Sir Isaac Newton's mathematical principles of natural philosophy and his system of the world. Vol. 1. The motion of bodies*. Berkeley: University of California Press.

Casey, J. 1992. The principle of rigidification. *Archive for History of Exact Sciences*. 43: 3229–3383.

Chalmers, A. 2012. Intermediate causes and explanations: The key to understanding the scientific revolution. *Studies in History and Philosophy of Science*. 43: 551–562.

Damerow, P., G. Freudenthal, P. Mclaughlin, and J. Renn. 2004. *Exploring the limits of pre-classical mechanics: A study of conceptual developments in early modern science*. New York: Springer.

Dobbs, B.J.T. 1991. *The Janus faces of genius: The role of alchemy in Newton's thought*. Cambridge: Cambridge University Press.

Ducheyne, S. 2012. *The main business of natural philosophy: Isaac Newton's natural-philosophical methodology*. Dordrecht: Springer.

Euler, L. 1954. *Leonhardi Euleri opera sub auspiciis Societatis Scientiarum Naturalium, Series II, 12*. Zurich: Orell Füssli Turici.

Gabbey, A. 2011. The term *materia* in Newton and in the Newtonian tradition, In: *Materia. Atti del XII Colloquia Internazionali del Lessica Intellettuale Europeo (Roma 7–9 gennalo 2010,* eds. G. Delfina, and V. Marco. Florence: Leo S. Olschki.

Hall, R.A., and M.B. Hall. 1962. *Unpublished scientific papers of Isaac Newton*. Cambridge: Cambridge University Press.

Henry, J. 2011. Gravity and *De gravitatione*: The development of Newton's ideas on action at a distance. *Studies in History and Philosophy of Science*. 42: 11–27.

Herivel, J. 1965. *The background to Newton's 'Principia': A study of Newton's researches in the years 1664–1684*. Oxford: Oxford University Press.

Newton, I. 1999. In *The principia: Mathematical principles of natural philosophy*, ed. I.B. Cohen and A. Whitman. Berkeley: University of California Press.

Pascal, B. 1937. *The physical treatises of Pascal: The equilibrium of liquids and the weight of the mass of the air.* Trans. A.G.H. Spiers and I.H.B. Spiers. New York: Columbia University Press.

Ruffner, J.A. 2012. Newton's *De gravitatione*: A review and assessment'. *Archive for History of Exact Sciences*. 66: 241–264.

Shapiro, A. 1974. Light, pressure and rectilinear propagation: Descartes' celestial optics and Newton's hydrostatics. *Studies in History and Philosophy of Science*. 5: 239–296.

Truesdell, C. 1954. Rational fluid mechanics, 1687–1765 In: Euler L. *Leonhardi Euleri opera sub auspiicis Societatis seintiarum maturalium*. Series II, 12: ix–cxxv. Zurich: Orell Füssli Turici.

Turnbull, H.W. 1959. *The correspondence of Isaac Newton, Vol. 1*. Cambridge: Cambridge University Press.

Westfall, R.S. 1962. The foundations of Newton's philosophy of nature. *British Journal for the History of Science*. 1: 171–182.

Chapter 10
Fashioning a Novel Concept of Pressure: One Hundred Years

Abstract It is enlightening to view the history of hydrostatics in terms of the overcoming of obstacles, obstacles that can only be recognized as such in retrospect. There are three candidates for inclusion in a list of such obstacles. From Archimedes to Stevin it had been assumed that the postulates of a science needed to be sufficiently obvious and unproblematic to be granted at the outset. The need for a science to outstrip the known and uncover significant novelty had yet to be appreciated. Another obstacle was a preoccupation with weight as the natural cause of hydrostatic phenomena. Thirdly, there was an inadequate appreciation of the fact that the functioning of liquids as constraints of a mechanical system needed specification. The constraints operative in the case of simple mechanical machines, such as the rigidity of balance arms, were too obvious to warrant explicit mention and so rendered the need to specify them invisible. The path from Stevin to Newton can be read as a gradual transcendence of such barriers. The identification of the way in which liquids constrain mechanical systems by virtue of the fact that they are unable to resist deforming forces, identified and exploited by Newton, was far from obvious at the beginning of the seventeenth century and had nothing to do with weight. It required the construction of a technical concept of pressure.

10.1 Achieving Progress by Overcoming Obstacles

According to the historical story that I have told in the foregoing pages, progress in hydrostatics in the seventeenth century came about largely through the replacement of a common understanding of pressure by a technical sense. The latter sense is not particularly difficult to grasp, so much so that much of the history of hydrostatics that I hope my version will supersede has taken it for granted and read it into the historical texts, mistaking the common sense conception of pressure that occurs in them for the modern sense that came to replace it. To the extent that the modern notion of pressure can seem obvious to us, it is legitimate to ask why it did not appear so at the beginning of the seventeenth century. What obstacles stood in the way of recognising the obviousness of the modern concept of pressure? Looking to history for an answer to such a line of questioning leads to a view of scientific progress that portrays it as coming about by way of the overcoming of obstacles. Such a history is necessarily backward looking insofar as 'epistemological obstacles' can

A.F. Chalmers, *One Hundred Years of Pressure*, Archimedes 51,
DOI 10.1007/978-3-319-56529-3_10

only be identified and recognised as such in retrospect.[1] Otherwise they would not be obstacles to progress. Backward looking though it may be, a historical case needs to be made for the identification of such obstacles and for the claim that they did indeed function as obstacles.

We have seen that hydrostatics grew out of a science of simple machines inspired by and modelled on the work of Archimedes. While such a path provided hydrostatics with much of what it needed it also put some obstacles in the way of progress. One such obstacle was the presumption that the claims of a science can be developed as the consequence of postulates sufficiently unproblematic to be granted *at the outset*. While some common facts about the behaviour of liquids could be granted on the basis of common sense at the beginning of the seventeenth century these were not sufficient to make a science of hydrostatics possible. The concept of pressure that was to prove to be necessary for progress required conceptual moves beyond common knowledge of the time.

There were two other obstacles that needed to be transcended before statics could be extended to hydrostatics. The constraints that were operative in the functioning of simple machines were not identified as such on account of their obviousness. This camouflaged the need for a precise specification of the constraints on a mechanical system that arise from the presence of a body of water. Closely connected with this is a second obstacle that arose from a preoccupation with weight as a natural cause of the functioning of machines. Such a preoccupation was present in hydrostatic theories from Archimedes to Simon Stevin and beyond and disguised the need for an adequate theorization of the fluidity of liquids.

In Sect. 10.2 I elaborate on the theme of hydrostatics as an extension of the science of simple machines and in Sect. 10.3 I recapitulate the one hundred year history of the transformation of the concept of pressure. I do so with a good deal of repetition, but abstract from the historical documentation provided earlier in the book. To some extent I view the past from the point of view of the present, but in a way that is intended to illuminate the kind of progress involved in the transformation of hydrostatics in the seventeenth century.

10.2 Hydrostatics as an Extension of Statics

When Simon Stevin set about formulating the foundations of hydrostatics in 1586 he had at his disposal a mathematized science of simple machines, to the development of which he himself contributed. He understandably saw his task as that of extending the science of simple machines to include hydrostatics. He not only presupposed the content of the science of weight but also adopted its style. Propositions of hydrostatics were to be derived as the consequences of postulates that could be

[1] A portrayal of progress in science via the overcoming of 'epistemological obstacles' that can only be recognised as such in retrospect was developed by Gaston Bachelard. See especially Bachelard (2002), an English translation of *La Formation de L'Esprit Scientifique* of 1918.

accepted as unproblematic at the outset. The extent to which extending knowledge into new areas would require the fashioning of novel concepts up to the task was not appreciated. In this respect, the adoption of the Euclidean ideal that was part of Stevin's Archimedean inheritance constituted an obstacle to progress.

Another obstacle involved a preoccupation with weight as the cause of hydrostatic phenomena. This was understandable insofar as weight was manifestly a cause of the phenomena with which the seventeenth century theorists were to grapple. Water is driven through conduits serving a town by the weight of the water in the reservoir; the ability of a liquid to support floating is dependent on its density; water 'weighs' against a lock gate. At least in retrospect it can be appreciated that a focus on weight was not destined to help with the task of developing a theory of phenomena distinctive of liquids since solids and liquids alike possess weight. Some theoretical grasp of the fluidity possessed by liquids but not by solids was needed. An appropriate notion of fluidity needed to be added to weight for the comprehension of hydrostatics. This proved to be a task that was far from straightforward.

Yet another obstacle stemmed from the role played by constraints in the understanding of simple machines. Material constraints played a crucial role in the functioning of machines that was implicit in the practice of artisans as well as those seeking to theorize statics. They serve to offer an explanation of the connection between cause and effect operative in those machines. The motion downwards of one weight on a balance necessarily results in an upward motion of the counterweight just because the connecting arm is rigid and constrained by the pivot. A falling weight is able to drive the hands of a clock because of the way in which the two are connected by gear wheels. Explanations of this kind, which I have chosen to refer to as mechanistic explanations, are intelligible in a common sense and constitute the kind of knowledge that a practitioner needs. They played a prominent part in the science of simple machines also, but were not explicitly discussed as such. The assumed naturalness and unproblematic nature of the functioning of mechanical constraints and the mechanistic explanations that they made possible at the turn of the sixteenth century are evident from the lack of explicit discussion of them. They were taken for granted.

To what extent were the constraints imposed on a material system by a body of liquid understood in a way that facilitated mechanistic explanations of the relevant phenomena? How are liquids able to push around corners? We have seen in the foregoing study that the constraints operative in hydrostatic 'machines' became available only when an account could be given of the way in which forces are transmitted from one point to another through the body of a liquid. This accomplishment required a technical notion of pressure that differed significantly from the common one.

My main argument for the appropriateness of considering hydrostatics in terms of the extent to which it provided mechanistic explanations lies in the fact that the acceptability of such explanations was taken for granted by practitioners and within the science of the simple machines. But there are other arguments for my position. As we have seen, both Isaac Beeckman and Robert Boyle had problems with

Stevin's version of hydrostatics that are readily translatable into the complaint that it failed to provide mechanistic explanations of hydrostatics. When Beeckman and Rene Descartes discussed some of the counter-intuitive claims of Stevin's theory that he had derived mathematically rather than explained mechanistically they responded by attempting to explain them in terms of corpuscle pushing against corpuscle. Boyle, for his part, complained that Stevin had merely shown that certain propositions are true without showing why they are true, and responded by elaborating on the concept of pressure. When Isaac Newton formulated hydrostatics by forging a concept of pressure very close to the one accepted today he completed the move to a mechanistic account of it by identifying how pushes are transmitted through the body of a liquid. It is of some significance that no need has been found to significantly modify Newton's theory to this day. Mechanistic explanations remain intelligible and useful when they can be had.

The claim in the foregoing sentence should not be taken to imply that mechanistic explanations that identify the way in which the parts of systems interact by way of contact action are essential to a science. By the end of the seventeenth century explanations that appealed to gravitational attraction were powerful but not mechanistic. A search for mechanistic explanations was not much help in the chemistry of the time. Mechanism proved to have its limits. Nevertheless, I hope I have demonstrated, both in the foregoing paragraphs and in the book as a whole, that it is apposite and instructive to view seventeenth-century hydrostatics as a quest for mechanistic explanations of a kind at work in the deployment and theory of simple machines.

The historical case for my standpoint does not end there. Around the middle of the seventeenth century the nature and status of mechanism and mechanical explanations became an explicit focus of attention in the context of the rise of the mechanical philosophy. The issues at stake need to be spelt out with some care because the notion of 'mechanical' as appropriated by the mechanical philosophers was not the same as the traditional notion and its association with the practice of mechanics.

What became referred to as the mechanical philosophy from the 1660s on was essentially an ontological thesis about the ultimate structure of material reality. That reality was presumed to be composed of corpuscles of universal matter each possessing a shape, size and degree of motion and nothing else. The mechanical philosophy was argued for mainly on the grounds of its intelligibility, a position that had force to the extent that explanations that involved corpuscles pushing against and colliding with each other were as straightforward as the explanations of the workings of simple machines. The mechanical philosophers contrasted this degree of intelligibility with the appeal made by scholastic Aristotelians to substantial forms and real qualities as features of the ultimate structure of reality. The mechanical philosophers considered those latter features to be unintelligible.

There are a number of things to be said about the notion of intelligibility as invoked by the mechanical philosophers. Firstly, the notion that inter-corpuscular interactions are intelligible traded on the sense in which the simple machines are intelligible and so took that latter intelligibility for granted. We saw in Chap. 8 that Boyle justified his use of the term 'mechanical philosophy' precisely because of the extent to which the kind of explanations it invoked 'are obvious and very powerful in mechanical engines' (Boyle 1999, Vol. 2, p. 87). That is, a basic assumption made by the mechanical philosophers was that mechanistic explanations are intelligible. Secondly, there are grounds for doubting that the likeness between the corpuscular mechanisms of the mechanical philosophers and simple machines manipulated by mechanics is sufficient to warrant the judgment that they are intelligible in the same sense. Important here is the point that the mechanical philosophy was a thesis about the ultimate structure of reality. The mechanical philosophers could hardly say that talk of weight is unintelligible, but they could more plausibly claim it unintelligible to treat weight as a property of corpuscles not subject to explanation at a deeper level. (In what direction would such a weight act?) Given this, there is a dis-analogy between corpuscular mechanisms and simple machines insofar as the former cannot presume such things as weight, rigidity and elasticity, all of which are involved in the comprehension of simple machines. There is also the problem that collisions between corpuscles, fundamental to the mechanical philosophy, threaten to be unintelligible insofar as they must take place instantaneously given that the corpuscles have an unchangeable shape and size. At the instant of impact, a corpuscle moves in two directions at the same time!

Boyle was forced to come to terms with the relationship between the two senses of 'mechanical'. This was because he was both an advocate of the ontological thesis involved in the mechanical philosophy and an experimenter who sought to extend the science of simple machines to pneumatics and hydrostatics. Insofar as those latter sciences were mechanical they were deemed to be so 'in the usual sense of that expression' (Boyle 1999, Vol. 7, p. 148). Accordingly Boyle was able to appeal to the spring of the air in his pneumatics, fluidity and pressure in his hydrostatics and weight in both of them. This was all without the need to reduce those entities to the shapes, sizes and motions of corpuscles which, as a committed mechanical philosopher, Boyle believed to constitute the ultimate components of the material world. Weight, spring and pressure constituted what Boyle occasionally referred to as intermediate causes to distinguish them from ultimate corpuscular causes. In articulating this stand Boyle was making explicit what was implicit in the approach of most of the main actors in our story. With the exception of Descartes and Beeckman, they all took weight for granted without any attempt to explain it at a deeper level. When pressure and spring entered into their deliberations they attempted to specify and deploy them but did not acknowledge a need to explain them by appeal to underlying corpuscles.

Seventeenth-century hydrostatics arose as an extension of the statics of simple machines to systems involving bodies of liquid. This required that the constraints imposed on a mechanical system by a body of liquid be specified. We have seen in the foregoing history that this involved the fashioning of a technical sense of pressure that differed from the common sense in use at the beginning of the seventeenth century. In the following section I review the contributions of the chief figures in our story from this point of view, aiming to capture the gist of the story unencumbered by detailed historical documentation.

10.3 The Path to a Concept of Pressure Adequate for Hydrostatics

The Archimedean account of floating was based on a balance between the weight of a solid and the upwards thrust of a liquid in which it is immersed. A common-sense distinction between liquids and solids was presupposed in the recognition that a liquid does indeed yield to make way for a solid weighing down on it. Liquids flow. A portion of liquid will move relative to neighboring portions if it is pressed more than those neighboring portions. The expression of this fact was the main postulate on which Archimedes based his theory, and it could be presented as grantable because it was presupposed in common knowledge of liquids.

Archimedes' theory was intended to deal with floating and as such it served his purpose well enough. Indeed, the basic account of floatation that it offered has persisted to the present day. However, the theory was ill equipped to deal with a range of hydrostatic phenomena. The assumption that a horizontal surface is pressed by, and only by, the weight of the water that lies vertically above it, while a natural one from Archimedes' perspective, is false. Further, his theory was quite incapable of dealing with hydrostatic forces other than vertical ones. The transformation of the downwards thrust of displaced liquid into an upwards thrust on an immersed solid implies that the pressing involved undergoes a one hundred and eighty degree turn, but this was given no attention in the Archimedean tradition. It was left in the domain of common sense.

One permanent legacy of the approach evident in Archimedes' treatment of floatation is the degree to which it involved abstraction and idealization. When Archimedes focused on water as such at the beginning of *On Floating Bodies* it served his purpose, not only to abstract from all other properties of water other than its weight and its capacity to flow, but also to remove solids from the picture as well. Starting, in thought with a world consisting solely of a sphere of water all parts of which weigh towards the center, Archimedes established that it would rest at equilibrium with its surface equidistant from the center. He then imagined the replacement of segments of cones of water by solids, themselves represented by their weight, solidity and dimensions, with all other properties ignored. It was in such terms that Archimedes arrived at his theory of floating. The lasting value of this strategy of abstracting and idealizing is evident from the fact that when Newton turned his attention to hydrostatics he too abstracted and idealized in a very similar way. He too began with a sphere of water, but abstracted to a degree that went even further than Archimedes had. Newton began with a liquid characterized simply by its ability to flow, ignoring its weight as well as all other properties. It is significant to note, at this point, that had Archimedes abstracted from weight he would have had nothing of significance left to work with!

When Stevin undertook to go beyond Archimedes to present a general theory of hydrostatics he inherited most of the features of his approach. The postulate that was to provide the crucial hydrostatic content of his theory expressed the commonly appreciated fact that the depth to which a vessel sinks is dependent on the amount of weight it carries. Neither in his definitions nor in his postulates did Stevin give formal expression to a feature of liquids that distinguishes them from solids. The capacity of liquids to flow was taken for granted in an unrefined common understanding. Because he followed the Euclidean ideal that he had inherited from Archimedes, the only way of developing a formal mathematized theory known to him, Stevin needed to base his theory on postulates that could be granted at the outset. As we now know, and as I have stressed in the foregoing history, an understanding of fluidity that makes it possible to understand the way in which a body of liquid constrains a mechanical system was not part of common knowledge near the end of the sixteenth century. Postulates adequate for a theory of hydrostatics needed to transcend what could be readily granted at that time.

As was considered in some detail in Chap. 3, Stevin's postulates, with their focus on weight and lack of a formal characterization of the liquid state, were incapable of yielding the content of his own theory. That content, which was innovative insofar as it incorporated and correctly evaluated the hydrostatic forces on planes other than horizontal, was introduced into his theory as additional, unheralded, and from the Euclidean point of view unwarranted, assumptions. These presumably had their origins in Stevin's experience as a hydraulic engineer.

Apart from the formal, mathematical, shortcoming highlighted above, Stevin's hydrostatics suffered from another, physical, one. It did not supply a complete account of the pushes responsible for hydrostatic effects. Stevin did not give a complete mechanistic account of hydrostatics. This was inevitably the case because Stevin's postulates were restricted to claims about weight, reinforced by some common knowledge about the capacity of liquids to flow. As was the case with Archimedes, Stevin's theory implied that hydrostatic forces were able to bend around corners, but there was no attempt to focus on, elaborate, or even to explicitly acknowledge this. Stevin was able to talk of the force on variously inclined planes using the term 'pressure' (*drucking*) and a range of other common terms. Insofar as there was a concept of pressure at work here it involved a liquid pressing against a solid surface. That is, it was pressure in a common sense with a long history.

A quarter of a century after Stevin had composed *The Elements of Hydrostatics* Galileo contributed to hydrostatics with an attempt to improve on the Archimedean account of floatation. As we saw in Chap. 4, he explicitly drew on the analogy between floating and the balance, showing how the balance conditions for the floating of a solid in a vessel could be derived by considering it as a balance with unequal arms and applying the inverse proportionality principle. This theory had the drawback that the explanation of any particular instance of floating requires that the vessel in which the floating takes place be implicated in the explanation. However, a more serious drawback lay in the fact that, while Galileo's theory enables the equilibrium conditions for any instance of floating, plus a few other states of hydrostatic equilibrium, to be deduced, it does not give a physical account, mechanistic or otherwise, of how the effects are brought about. Once again, the fact that liquids freely give way to the passage of solids is taken for granted, but no explicit attention is given to the implications this has for the way in which hydrostatic forces are communicated through liquids. The explicit way in which Galileo's hydrostatics invoked an analogy with weights on a balance ensured that the theory was incapable of coping with hydrostatic forces other than vertically acting ones. Had Galileo been concerned with developing a theory of hydrostatics his account of floating would not have provided him with much of what was needed.

When Beeckman expressed the view that 'Stevin was too devoted to mathematics and dealt too rarely with physics' he had Stevin's work on music in mind, but might well have directed it at his hydrostatics.[2] When he confronted Descartes with some puzzling claims from Stevin's hydrostatics it was clear that what he sought and what he hoped Descartes would provide was what I have called mechanistic explanations of them. It is equally clear that what they both aspired to were mechanistic explanations that involved corpuscle pushing against corpuscle. The precise way in which Descartes pursued that idea through his subsequent deliberations traced hydrostatic forces back to the weights of corpuscles which were themselves attributed to the centrifugal forces radiating out from the axis of the minor vortex of second element circulating the earth. Insofar as hydrostatic forces were pressures for Descartes they were linear pressures arising from corpuscle pressing on corpuscle. This led Descartes to a hydrostatics that was in conflict with a range of hydrostatic phenomena and which was quite incapable of coping with the isotropy exhibited by many of them.

The first significant moves from a common towards a technical sense of pressure were made by Blaise Pascal. He recognized that hydrostatics needed a grasp of the way in which liquids differed from solids and located those properties in their 'continuity' and 'fluidity'. The details of what he intended here were spelt out by reference to an idealized version of what we would call the hydraulic press. Continuity is taken to imply that a piston in one aperture of the press cannot be moved without moving a piston in the other, where it is clear that continuity included incompressibility as well as spatial continuity. Fluidity is taken by Pascal to mean that a pressing exerted at one location in the boundary of an enclosed liquid, evaluated as a force per unit area, will be transmitted to any other part of its boundary, whatever its orientation, as the same force per unit area. Consequently, a device such as the hydraulic press is a mechanical machine for multiplying force. The fact that the operation of hydraulic machines ceases if the liquid communicating the pressing is frozen demonstrates that the hydrostatic effects in evidence here are due to the specific way in which liquids, as opposed to solids, transmit forces applied to them.

In his deliberations, Pascal took a step towards separating forces due to pressure from those due to weight. The force pressing on an aperture of the hydraulic press can arise in a number of ways besides the action of a weight. However, whatever the cause of that force it will be transmitted to other parts of the container in the way specified by Pascal. In considering the force per unit area at a location at a different height than the aperture at which the force is applied the effect of the weight of the liquid in the body of the press will need to be taken into consideration. The force arising from the applied pressure is distinct from that arising from the weight of the liquid that transmits it.

[2] See de Ward (1939–1943), Vol. 3, p. 52.

It is noteworthy that in his exposition of how his account of the transmission of pressure can explain a number of hydrostatic effects, Pascal identified the pushes that give rise to them. That is, he offered mechanistic explanations. For instance, when he explained the apparent loss in weight of a body immersed in a liquid, Pascal attributed it to the difference between the extent to which the liquid presses on the top and the base of the body, rather than by invoking the weight of liquid displaced and employing an analogy with the balance. Mechanistic explanations such as this remained in the domain of the common meaning of pressure insofar as they related to forces on a solid surface bounding a liquid. As was the case with the theories of his predecessors, the precise way in which pressures are transmitted through the body of a liquid remained unspecified and unexplored. Pascal did not make the move to a conception of pressure as representing a state within the body of a liquid, a move that requires that pressure be understood as a scalar rather than vector quantity.

When Boyle turned to Pascal's *Equilibrium of Liquids* in 1664 he had already made his experimental and theoretical innovations in pneumatics. He had moved beyond weight by attributing a spring to air, distinct from weight and representing a state possessed by a sample of air throughout its volume analogous to the way a stretched string possesses a tension throughout its length. The centrality of spring in Boyle's account of the behavior of air, as opposed to solids, must have alerted him to the issue of how liquids are to be distinguished from solids, since they lack a significant spring. Boyle was able to build on Pascal's hydrostatics to do this. He did so by way of a theoretical device that was innovative and crucial in spite of its obviousness from a modern point of view. Boyle considered forces acting on either side of imaginary planes within the body of liquids in equilibrium. He exploited this device to represent the fact that the force per unit area on a plane within a liquid depends only on its depth and is independent of its orientation. In this way, Boyle went some way towards giving a mechanistic understanding of the transmission of hydrostatic forces through liquids. Boyle did not make explicit the point that pressure within the body of a liquid tends equally in all directions resulting in no net force. Further, insofar as he focused on pressure at a location within a liquid as dependent on the depth of that location beneath the liquid surface, he did not completely dissociate weight and pressure. Boyle did make a step in that direction, however, when dealing with the fact that a solid body can float in a body of water possessing a weight much less than its own.

By building on the hydrostatic tradition summarized above, Newton was able to construct a version of hydrostatics employing a technical concept of pressure that conforms closely to the modern one. By abstracting and idealizing in a way analogous to the strategies of Archimedes and Stevin, Newton was able to present his theory in a thoroughly mathematical form. His first step assumed a sphere of incompressible liquid devoid of weight pressed equally at all locations on its outer surface. He then proceeded to explore the pressure on arbitrary planes within the liquid, taking for granted, and without comment, the innovation introduced by Boyle. As far as capturing the way in which liquids differ from solids is concerned, Newton recognized that their fluidity, as assumed by Pascal and Boyle and implicit in common knowledge, could best be formulated as their inability to resist distorting forces. He recognized that it was necessary to assume that fluids are strictly continuous, any portion of liquid, however small, possessing that property. Only in this way could Newton's theory grasp the way in which fluids transform directed forces into isotropic ones in a way that solids and powders do not. Continuity was assumed in Newton's mathematical proofs insofar as they involved spheres touching planes and other spheres at points. Newton explicitly acknowledged that within a body of liquid in equilibrium pressure acts equally in all directions. His exposition of pressure clearly established it as a property distinct from weight, which Newton subsequently added, as a second step, in the version of his theory presented in the *Principia*.

I may have been over generous in reading into Newton's definition of a fluid all of the import of the modern conception. It was Leonhardi Euler, in 1753, in his 'General Principles of the Equilibrium of Fluids' who made it quite explicit that the defining characteristic of a fluid is that the force on any surface within it must be normal to that surface if the fluid is to be in equilibrium, with the consequence that such a fluid is free of shear stresses (Euler 1954, p. 3). Bringing what Clifford Truesdell (1954, p. x) described as his constant quest for 'first principles, generality, order and, above all, clarity' to bear on hydrostatics, Euler exploited mathematics of the continuum that had advanced beyond what Newton had at his disposal to develop a formal treatment of the statics of fluids, which corresponds pretty much with what is accepted today. He generalized the theory to include compressible fluids, that is, air, so that hydrostatics became a special case of fluid statics. It was Euler who introduced the mathematical device of conceiving of elements of surfaces and integrating the forces on them, thereby formalizing the notion that I used when I, perhaps over-charitably and anachronistically, summarized the content of Stevin's law for the benefit of the modern reader.

I refrain from exploring the historical details of Euler's contributions. After Newton the development of hydrostatics became inextricably engaged with hydrodynamics, and exploring that would take me too far afield from the main points I wish to illustrate with this history. As a matter of fact, it was in fluid dynamics that the notion of pressure became absolutely crucial. As far as hydrostatics is concerned, forces are manifest only as the forces on the solid surfaces that bound liquids, the pressures throughout the interior of liquids cancelling each other out. Given this, one might well take Pascal's representation of the forces on solid surfaces as sufficient and Boyle's addition of forces in the body of liquids as superfluous. Such is no longer the case in hydrodynamics, where the variation of pressure from point to point is a crucial determinant of the resultant motions.

We have seen that the move from a common understanding and use of pressure, from the perspective of which directed pressures are paradigmatic rather than meaningless, to the modern, technical concept of pressure, was far from obvious or straightforward. It took most of the seventeenth century for the change to be accomplished. The technical notion, taken for granted in contemporary physics and engineering and playing a crucial role in the practical advances of the Industrial Revolution, became available only after it had been made obvious and straightforward as a result of the theoretical initiatives of the likes of Stevin, Pascal, Boyle and Newton.

In this section I have summarized the seventeenth-century path to a theorized hydrostatics via the evolution of a formal concept of pressure that was up to the task. The formulation and deployment of the new concept was essentially mathematical in a way that was made possible by abstraction and idealization of a kind that dated back to Euclid and Archimedes. Stevin's deployment of 'surface vessels', Pascal's articulation of his theory by reference to an idealized hydraulic press constructed in thought, Boyle's conception of forces acting on imaginary planes within the body of a liquid and Newton's analysis of a uniformly compressed sphere of water and sub-spheres within it touching each other at points, all attest to this fact. What the summary presented here leaves untouched is the relationship between the various versions of hydrostatics and knowledge of the phenomena they were intended to explain. In what sense was hydrostatics an empirical or experimental, as well as a mathematical, science? I review the answers to this question that can be gleaned from the foregoing history of hydrostatics in the next chapter.

References

Bachelard, G. 2002. *The formation of the scientific mind*. Trans. M. McAllester. Manchester: Clinamen Press.

Boyle, R. 1999. In *The works of Robert Boyle*, ed. M. Hunter and E. Davis, vol. 14. London: Pickering and Chatto.

de Ward, C. 1939–1935. *Journal tenu par Isaac Beeckman de 1604 de 1634*. Vol. 4. The Hague: Nijhoff.

Euler, L. 1954. *Leonhardi Euleri opera sub auspiciis Societatis Scientiarum Naturalium*, Series II. Vol. 12. Zurich: Orell Füssli Turici.

Truesdell, C. 1954. Rational fluid mechanics, 1687–1765. In *Euler L. Leonhardi Euleri opera sub auspiicis Societatis seintiarum maturalium*, Series II. Vol. 12, ix–cxxv. Zurich: Orell Füssli Turici.

as a result of the theoretical initiatives of the likes of Stevin, Pascal, [illegible]

In this section I have summarized the seventeenth-century [illegible]

[illegible]

Chapter 11
Hydrostatics and Experiment

Abstract This chapter is concerned with capturing exactly what is involved in the idea that the progression in hydrostatics from Stevin to Newton corresponded to the shift from the attempt to construct a science on the basis of given, unproblematic, postulates to the recognition that the adequacy of postulates needed to be discovered and justified via experiment. The importance that experiment could play in the discovery of previously unknown phenomena was clear in the context of pneumatics in the wake of Torricelli's experiment. However, as far as hydrostatics is concerned, the move from Stevin to Newton did not involve discovery of hydrostatic phenomena that were significantly novel. If a theory is supported by a range of evidence to the extent that it can explain it in an uncontrived way, then it can be supported by old as well as new evidence. The path from Stevin to Newton can be seen as involving steadily increasing degrees of support from a largely unchanging pool of evidence because of the extent to which that evidence could be naturally explained to an increasing degree. Other roles for experiment in seventeenth-century hydrostatics can be clearly discerned in the work of Boyle. Those roles concern the way in which thought-directed manipulation can contribute to novel concept formation and the way in which the exposition of a theory can be facilitated by key thought experiments or archetypal experiments actually performed.

11.1 Experimental Philosophy Versus the Euclidean Ideal

In the previous chapter I reflected on the changes in the concept of pressure that were involved in progress within seventeenth-century hydrostatics. I noted how this clashed with the notion, implicit in what I have called the Euclidean ideal, that the postulates of a science must be sufficiently unproblematic to be granted at the outset of its development. Concepts adequate for a science need to be constructed rather than drawn from what is already to hand. There was another, equally pressing, need to move beyond the Euclidean ideal. In the course of the seventeenth century it became increasingly the case that key phenomena relevant for a science were not common knowledge. Rather, they needed to be discovered by way of experimental exploration, and the acceptability of the axioms of a science was to be judged by the extent to which they were borne out by observation and experiment. Such features lay behind the appearance of the term 'experimental philosophy' to describe the

A.F. Chalmers, *One Hundred Years of Pressure*, Archimedes 51,
DOI 10.1007/978-3-319-56529-3_11

new sciences from the 1760s onwards. From this point of view, experimental practice played a vital epistemological function that went way beyond mere application of knowledge established independently of it, an assumption that formed part of the adoption of the Euclidean ideal by Simon Stevin.

The above assertions must surely be true of seventeenth-century developments in science in at least a rough and ready sense. However, in this chapter I aim for more precision, and more enlightenment, by pursuing the issue in greater depth in the context of our history of hydrostatics. Two features of the history of hydrostatics would seem to pose problems for the view that seventeenth-century science involved a move from Euclidean to experimental science. First, all of the phenomena explained by Isaac Newton's hydrostatics towards the end of the seventeenth century were known to, and exploited by, Stevin before that century began. It is true that Stevin departed from the Euclidean ideal by incorporating into his theory empirical knowledge not included in his postulates, but that additional knowledge was drawn from practical knowledge familiar to practitioners such as he was and did not rely significantly on artificial experimentation. Advances in seventeenth-century hydrostatics did not rely on appeals to novel observations and experiments as providing evidence for it. Second, we have seen that when Newton formulated the version of hydrostatics that represents the culmination of the quest for a theory of hydrostatics described in the foregoing pages, he represented it as following from the definition of a liquid. And it was only after Newton had proved the propositions containing the substance of his theory that he showed how they could be used to explain a range of hydrostatic phenomena. He, in effect, represented his hydrostatics as conforming to the Euclidean ideal!

In this chapter I seek to clarify the sense in which seventeenth-century 'experimental philosophy' was empirical science borne out by observation and experiment as opposed to something more akin to mathematics as exemplified by Euclidean geometry. I will do so by reflecting on the roles played by observation and experiment in hydrostatics.

11.2 The Practical Applicability of Euclidean Geometry and Statics

Questions concerning the practical applicability of mathematically formulated theories can be raised in the context of Euclidean geometry itself. How does it come about that geometrical theorems derived from abstract definitions of 'straight line' and the like apply to the real world to the extent that its theorems, like the one named after Pythagoras, can be confidently assumed to be true independent of any measurements or applications? The answer, surely, lies in the fact that definitions of and propositions concerning entities such as point and straight line are not arbitrary or merely conventional but are abstractions from real world situations that were encountered and dealt with long before Euclid's formulations. The notion of a straight line is presupposed in the practice of pointing to indicate direction, in the recognition that the most efficient way to get from one location to another is to proceed as straight as the terrain permits and in the appreciation that general features of shadows arise because light is transmitted linearly. Such knowledge was presupposed and articulated in surveying, architecture and astronomy before the formalization of geometry by Euclid. A formal notion of a straight line involved in geometry is an abstraction of what is instantiated in stretched strings, plumb lines and lines of sight, knowledge that is already general because of the multiple instantiations of the notion of a line.

Geometrical propositions about figures bounded by straight lines are approximately true of material triangles and the like because the concepts involved and their implications are, at least initially, abstracted, tidied up versions of what is encountered in everyday dealings with and knowledge of the world. We are now able to appreciate the fact that when dealings with the everyday world are transcended, as it is in large-scale astronomy, then that geometry ceases to apply, the validity of its proofs notwithstanding.

When Stevin constructed his geometrically formulated account of simple machines in *The Art of Weighing*, building on the earlier work of Archimedes, he needed to include in his definitions and postulates assumptions that took him beyond geometry. As far as the static equilibrium addressed in his theory is concerned, propositions sufficient for his purpose could be abstracted from common knowledge of his day in much the same way as the notion of a straight line could. Material objects have a weight that bears them down towards the center of the earth. If one such object is added to another their combined weight is equal to the sum of their individual weights. If one weight is joined to another via a rigid rod supported by a pivot, then the downwards displacement of one weight will necessitate the upwards displacement of the other. Equilibrium will occur if equal weights are supported on equal arms. These are unproblematic claims abstracted from common knowledge and implicit in the age-old use of balances. The mathematical formulations developed by Stevin concerned the centers of gravity and equilibrium of objects and arrangements that have ideal geometrical forms, be they prisms, cones, spheres or whatever. Theorems concerning equilibrium arising from their weights will apply to material instantiations of such shapes to the extent that the latter approximate the ideal forms assumed in the geometrical proofs. The mathematical theory had empirical content because its definitions and postulates, sufficient to yield the relevant theorems as consequences, had links to common knowledge of the world, much of the latter presupposed in age-old technologies.

The picture I have portrayed in this section enables us to understand why there could be such a thing as mathematically formulated theories applicable to the observable, manipulable world well before the innovations of the seventeenth century. It also enables us to foreshadow the ways in which the procedures involved in that mathematical science would prove to have their limits. The procedures incorporated into the Euclidean ideal proved to be inadequate when everyday knowledge was transcended by way of recondite knowledge produced by novel experiments. Further, the conceptualisations necessary for the mathematical formalization of the new knowledge could not simply be read off of nature. They needed to be constructed. We have dealt in the previous chapter with the problem that an adequate conceptualisation of fluidity posed and how it was eventually met via a technical sense of pressure. The focus in this chapter is on the empirical basis of hydrostatics and its relation to experiment. To what extent was hydrostatics an early instantiation of the experimental philosophy?

11.3 Novelty via Experiment

To an increasing extent as the seventeenth century progressed, phenomena were revealed by the performance of contrived experiments. Knowledge acquired in this way necessarily transcended what was already known and so could not necessarily be accommodated into theoretical systems based on postulates that could be granted on the basis of what was ready to hand. In Chap. 7 I described how a science of pneumatics was initiated in response to a series of novel experimental discoveries starting with Evangelista Torricelli's in 1644 and taken much further by Robert Boyle's experiments with his air pump. Blaise Pascal actively participated in those advances, designing and sometimes performing experiments of his own. He came to the conclusion that 'experiment is the true master that one must follow in physics', arguing that the new experiments could be used to refute the age-old belief that nature opposes the formation of a vacuum and identifying the weight of air as the cause of pneumatic phenomena.[1]

The new emphasis on experiment was highlighted by Pascal in his treatise on hydrostatics from its opening page, as we have seen. In the next section we will see that there was some justification for such an emphasis. However, many of the phenomena appealed to and explained by Pascal in his hydrostatics were long-known phenomena such as floating and the hydrostatic paradox. When Pascal did invoke contrived experimental set-ups they were, in the main, versions of what could be found in Stevin's *Practice of Hydrostatics* and treated there as practical applications rather than as sources of evidence. The hydraulic press was an exception, but it is doubtful that Pascal ever constructed one. It is difficult to document with certainty when and by whom the experiments described by Stevin were actually performed but it is clear that they were treated as experimental facts by the seventeenth-century actors in our story.[2] To the extent that Pascal based his theory on observational and experimental evidence it was not novel observational and experimental evidence.

Reliance on knowledge of novel phenomena produced by contrived experiments was to become a notable feature of seventeenth century science and posed problems

[1] I noted in Chap. 7 that Pascal failed to appreciate the need to augment the weight of air with a property corresponding to what Boyle was later to call its spring as anticipated by Pascal's contemporaries Roberval and Pecquet.

[2] I defended this point in Sect. 6.5, invoking the treatment of Stevin's empirical claims by Mersenne, Beeckman and Descrates.

for the Euclidean ideal insofar as this was the case. Such a feature can confidently be included in what was intended by the phrase 'experimental philosophy'. Nevertheless, for the reasons I have summarised above, seventeenth-century hydrostatics did not significantly exemplify experimental philosophy *in that sense*. Whether there were other senses in which it did so qualify is an issue we pursue in the remainder of this chapter.

11.4 Experiments as Suppliers of Evidence

The evidence appealed to in support of theories of hydrostatics in the seventeenth century was not novel evidence. It was comprised of common knowledge, including that entailed in well-developed technologies, to which can be added the behaviour of the contrived situations described by Stevin in *The Practice of Hydrostatics* in 1586. In this section I explore the relationship between the strength of evidence, on the one hand, and its novelty on the other. I argue the case that evidential support can be strong without being novel and use the point to defend the claim that seventeenth-century hydrostatics came to be strongly supported by empirical evidence, including experimental evidence which was not novel.

Except for the occasional exception, the theories of hydrostatics that we have analysed in this book were not defended by their proponents by appeal to some explicitly articulated account of how theories can be supported by observation and experiment. The best I can do is to attempt to identify some of the intuitions that guided their deliberations. Some of them can be illustrated by reference to the case for the Copernican theory which all of the actors in our story accepted as adequate. Prior to Galileo's appeal to observations using a telescope, the case for the superiority of the Copernican theory over its rivals was based on knowledge of planetary motions that dated from the time of the Ancients. This included the retrograde motion of the planets and appreciation of the fact that unlike the other planets, Venus and Mercury as viewed from earth are never far from the Sun.[3] Phenomena such as these followed, at least qualitatively, in a straightforward way from the assumption that the earth, along with the other planets, orbits the Sun. By contrast, Earth-centred versions of the planetary system could accommodate the evidence only in a contrived way, by adding epicycles tailor-made for the purpose. Rene Descartes alluded to the intuition in operation here when, in a letter to Jean-Baptiste Morin in 1638 he observed that 'it is true that there are many effects to which it is easy to fit many separate causes, but it is not so easy to fit a single cause to many different effects, unless it is the true cause that produces them'.[4] A theory is

[3] The broad features of the case for the Copernican theory are well described in Thomas Kuhn's classic *The Copernican Revolution* (1959).

[4] Cottingham et al. (1991, p. 107). Towards the end of his *Principles of Philosophy* (Section 205) Descartes illustrated the moral to be drawn here with an analogy with code-cracking. Just as the fact that a number of encrypted messages can be transformed into intelligible ones is evidence that

supported by observation and experiment to the extent that it can explain a wide range of such evidence in an uncontrived way. The example of the Copernican theory reveals why the qualification 'uncontrived' is important. It would be a co-incidence if a theory were able to naturally fit a range of evidence if it were false, but no co-incidence if in each case the theory were modified or supplemented to match the evidence in an ad-hoc way with an eye to the phenomena that needed to be accommodated.

To the extent that this account of empirical support is acceptable, it makes no difference to the degree of support whether the evidence is available before or after the formulation of the theory. It is true that observation of the phases of Venus by Galileo added to the support of the Copernican theory. But it did so by virtue of the fact that those phases follow as a natural consequence of the theory (once it is accepted that Venus shines by way of reflected sunlight), not because it was novel.

If we adopt the above account as representing an intuition about empirical support accepted by the actors in our story, then we can look anew at the case that Pascal made for his hydrostatics. He began *The Equilibrium of Liquids* by presenting some experimental facts concerning liquids that were to some extent counter-intuitive but not new to hydrostatics. He then explained those facts by appeal to a theoretical proposition sometimes referred to as Pascal's principle. A force applied at one location at the solid surface bounding a liquid, measured as a force per unit area, will appear as the same force per unit area at any other location on the bounding surface. In the remainder of the book he showed how a wide range of other hydrostatic phenomena could be seen as a natural consequence of his principle. Many of those phenomena, such as floating and the ability of water to find its own height, had been long known. Others involved contrived experimental situations similar to those appearing in Stevin's *Practice of Hydrostatics*. By the end of the book he had made a strong empirical case for his theory.

Pascal's reasoning so summarised did not involve novel experimental discoveries, notwithstanding the misleading impression conveyed by Pascal that he had conducted a range of experiments in support of his hydrostatics. Nevertheless, it marked a significant departure from the Euclidean ideal that Stevin had aspired to. In particular, Pascal's principle was not presented as an unproblematic postulate to be accepted at the outset. Rather it was one that was to be accepted on account of the breadth of observational and experimental evidence that it could straightforwardly explain, a mode of argument that does not require that the phenomena appealed to be novel. This use of observation and experiment as evidence can reasonably be included in what seventeenth-century usage of the phrase 'experimental philosophy' was intended to capture.

So far, so good, but we have yet to face up to the complication of such a picture by Newton's characterisation of his hydrostatics, especially in *De gravitatione*, as conforming to something that reads very much like the Euclidean ideal.

---

the correct cipher has been found, so the fact that a theory can straightforwardly explain a wide range of phenomena is evidence for its truth.

11.5 Newton's Hydrostatics and the Euclidean Ideal

We saw in Chap. 6 how Newton introduced *De gravitatione* as a work on hydrostatics and, as an opening gambit, distinguished between two methods. One was the mathematical method in which propositions are deduced from principles abstracted from physical situations and well known to the student. The other 'freer' method involved demonstrating the usefulness of a doctrine and 'perhaps confirming it' by appeal to experiment. Newton made it clear that his hydrostatics was to follow the former of these two methods. He set out to derive his theory of hydrostatics from 'definitions of certain words' and 'axioms and postulates denied by none'. Newton's characterisation of his enterprise fits closely with what I have termed 'the Euclidean ideal'.

Important amongst the ensuing list of definitions listed by Newton in *De gravitatione* was his definition of a fluid. We have noted that Newton procrastinated over this definition. What is more, he did not use it in his subsequent proof of the key proposition that the parts of a non-gravitating fluid in equilibrium press each other equally on all sides, with the result that his proof was invalid. How to define a fluid and use it to formulate propositions 'denied by none' that were sufficient to form the basis of hydrostatics was not perfectly clear to Newton himself at the time, let alone the 'student'. We do not know how Newton intended to complete his hydrostatics in *De gravitatione* by re-introducing the gravity from which he had abstracted and perhaps by commenting on the links between his theory and hydrostatic phenomena. Newton's tract remained incomplete.

As far as the substance of Newton's hydrostatics is concerned, the version in the *Principia* can be read as an improvement on and completion of the earlier version. The initial definition is replaced by a more adequate one explicitly appealed to by Newton in his proofs, the theory is extended to include gravitation and the ability of the theory to explain a range of phenomena, including floating and the pressure exerted on the base of various tortuous tubes of varying shapes and sizes is demonstrated. What is not included in the more mature version is the meta-discussion of the theory present in *De gravitatione*. In particular, Newton did not invoke his distinction between the two methods that he stressed in the opening of the earlier work.

A theme that was stressed by Newton in the *Principia*, from the first edition onwards, is that knowledge of the world is to be borne out by phenomena. A particularly forthright statement of that view appears in an unpublished version of a Preface composed around the time of the publication of the second edition in 1713.

> In all philosophy we must begin from phenomena and admit no principles of things, no causes, no explanations, except those which are established through phenomena.[5]

---

[5] Newton (1999, p. 54). In the Preface to the First Edition of the *Principia* Newton described the 'basic problem' in philosophy to be 'to discover the forces of nature from the phenomena of motions and then to demonstrate the other phenomena from those motions' Newton (1999, p.182). The fourth of the 'Rules of Reasoning' with which Newton began Book 3 of the *Principia* expressed the view that in 'experimental philosophy' propositions established by induction from phenomena are to be treated as true until the discovery of further phenomena should require that

The details of how to make significant and precise sense of such avowals in the context of the *Principia* have been given much attention in the literature.[6] Whatever those details, it is clear that the position that the mature Newton took on the need to base knowledge on the phenomena clashes with the view that it should conform to the Euclidean ideal. In particular, it clashes with his presentation of hydrostatics in *De gravitatione* as following what he called the mathematical method. It is not difficult to see how Newton could have represented the version of hydrostatics that he presented in the *Principia* as 'established through phenomena' had he taken the time to do that. While it can be conceded that his theory follows as a consequence of his definition of a liquid, plus some commonly recognised features of gravitation, there remains the grounds on which his definition is to be accepted as correct and sufficient for his purpose. The idea that liquids, by virtue of being such, cannot withstand a distorting force can be rendered plausible by reference to common phenomena such as the spreading of liquids over a surface when free to do so and the degree to which they give way to solids moving slowly through them. However, the fact that this definition is sufficient to explain the range of hydrostatic phenomena, both common and experimental, known in the seventeenth century is not obvious and needs to be demonstrated. By providing the necessary derivations Newton in effect constructed an empirical case for the claim that his definition of the liquid state was correct and sufficient. It is significant, in this respect, that the move from the hydrostatics of *De gravitatione* to that of the *Principia* involved a modification of both the definition of a liquid and the ensuing proofs based on it. The adequacy of the definition was dependent on the range of phenomena that could be derived from it and thus explained by reference to it. For this reason Newton's hydrostatics can be seen as experimental as opposed to Euclidean science, in keeping with Newton's claims concerning the need to base knowledge on the phenomena.

Insofar as the mathematical method advocated by Newton in the hydrostatics of *De gravitatione* clashed with his insistence on the need to base knowledge on the phenomena in the *Principia*, there are grounds for believing that the former work was written by a less-mature Newton. This would have been prior to the development of his views on derivations from the phenomena which became crucial in the defence of his deployment of universal gravitation in his 'system of the world'. My reading of the case gives support to an early dating of the composition of *De gravitatione.*

---

they be revised or qualified. In the General Scholium added to the Second Edition Newton insisted that 'whatever is not deduced from the phenomena' must be regarded as hypotheses of a kind that have no place in 'experimental philosophy' in which 'propositions are deduced from the phenomena and are made general by induction (Newton 1999, p. 943).

[6] Recent works that illustrate the degree of subtlety involved are Smith (2002) and Harper (2009).

11.6 Additional Roles for Experiment

Experiment in Boyle's work certainly played the two roles attributed to experiment discussed so far in this Chapter. His experiments with the air pump provided classic examples of the experimental production of phenomena not otherwise known and he deployed them as providing evidence for his claims about the spring of the air. However, there were other roles played by experiment in Boyle's work that can be legitimately listed as valid components of the new 'experimental philosophy'. They concern the way in which experimental manipulation can help provide a path to conceptual novelty and the use of experiment in the exposition of a theory. I offer a brief discussion of these two aspects of experiment in this section in the context of hydrostatics.

Theoretical innovation and the conceptual means of expressing it can emerge in the context of practical manipulations of the world. Successful manipulation needs to be guided by thought. Conversely, the adequacy of the thought is revealed by the success of the manipulations that they encourage and make possible. A classic example of the integrated work of head and hand involved here is the way in which Michael Faraday's conception of electromagnetic phenomena in terms of lines of force and his practical ability to identify and exploit the phenomena evolved together. David Gooding (1990) has traced this evolution in Faraday's work, taking advantage of the detailed descriptions of the latter's day to day struggles recorded in notebooks. In Chap. 8 I suggested that Boyle's description of his experimenting in *Hydrostatical paradoxes* can be read in a similar way. Boyle adopted the procedure of exploring the direction and magnitude of the pressure at locations within a body of water by inserting a tube holding a column of oil or other liquid, the height of the column being indicative of the force acting on the oil in the lower end of the tube. I suggest that Boyle came to think of himself as using the tube to detect pressures pre-existing in the body of water and detected by means of his search tube. In this way he was led to the theoretical device of conceiving of pressure acting on either side of variously orientated planes within a body of water. My suggestion is motivated by two historical facts. First, the device of conceiving of pressure as acting on hypothetical planes, which is taken for granted by Newton as it is in modern science, was first given clear expression by Boyle. Second, it was Boyle who seems to be the first to have carried out the detailed exploration of pressure in water by the practical manipulations described in *Hydrostatical Paradoxes*, the very same work in which the theoretical innovation was introduced. Boyle made that innovation without fanfare and without foreseeing the highly significant way in which Newton was to exploit it.

Boyle downplayed the extent to which the experiments described in *Hydrostatical Paradoxes* served as evidence for hydrostatic theory. The reasons he gave for publishing that work in its Preface appealed to other factors. Those reasons were concerned with making the new hydrostatics available to a wide audience, including those not well versed in mathematics. Boyle expressed dissatisfaction with works on hydrostatics. He characterised some of them as inadequate because they were based on discredited Aristotelian assumptions and others because they were written as works in mathematics and, as such, failed to capture physical explanations of natural phenomena.

> Wherefore, though neither the occasion and design of this Treatise exacted, nor my want of skill and leisure qualified me to write, either a body or elements of hydrostatics; yet I hoped I might do something, both towards the illustrating, and towards the rescue of so valuable a discipline, by publishing the ensuing tract; where I endeavor to disprove the received errors, by establishing paradoxes contrary to them, and to make the truths the better understood and received, partly by a way of explicating them unemployed in hydrostatical books, and partly by confirming the things I deliver by physical and sensible experiments. And over and above this, the more to recommend hydrostatics themselves to the reader, I have, besides the paradoxes, opposed to the errors I would disprove, taken occasion by the same way, to make out some of the usefullest of those hydrostatical truths, that are wont to seem strange to beginners (Boyle 1999, Vol., p. 195).

The main role of Boyle's experiments was to function as a device for explaining and elucidating the kind of physical explanations afforded by the new hydrostatics. They were designed to make the theory intelligible to newcomers to hydrostatics and to those lacking in mathematical skills, both of which characteristics could be ascribed to some members of the Royal Society. Michael Hunter and Edward Davis describe the way in which *Hydrostatical Paradoxes* was well received at the time and 'added to Boyle's mounting reputation as the leading experimental philosopher of his day'.[7] Boyle's experiments in hydrostatics as he characterised them played a role similar to that played by exemplary experiments in modern textbooks. Pascal's use of the hydrostatic press can be cast in a similar light, in spite of the fact that he did not perform it.

Seventeenth-century experimental philosophy involved the discovery of novel phenomena by way of experiment, an increased reliance on artificial experiment as providing the evidence for theoretical claims, the emergence of new ways of thinking and doing by way of thought-directed manipulations of nature, and the use of paradigm experiments to explicate a theory. Our history of hydrostatics can be utilised to shed light on all four aspects of experiment as it was deployed in the seventeenth century.

11.7 Experimental Philosophy Versus the Mechanical Philosophy

Alongside emergence of the term 'experimental philosophy' to refer to seventeenth-century developments was the emergence of another term, 'mechanical philosophy'. That latter term was employed by the seventeenth-century actors from the 1660s onwards and subsequently by historians of science to refer to a new

[7] Hunter and Davis make their remarks in their introductory notes to *Hydostatical Paradoxes* in Boyle (1999, Vol. 5, p. xxii). They cite Samuel Pepys as an early beneficiary of Boyle's exposition of hydrostatics.

world-view developed as an alternative to the Aristotelian one. In this section I explore the relationship between experimental philosophy and the mechanical philosophy and argue that articulation and adoption of the latter was not a significant factor as far as the emergence of hydrostatics is concerned. It was too remote from what could be tested experimentally to perform such a function.

There is no doubt that, in general terms, the world was seen in a very different way by the end of the seventeenth century than it had been at the beginning. As far as articulations of the change by philosophers and theologians are concerned there is rough validity in the idea that an Aristotelian world-view was replaced by a mechanical one. The century in which modern science emerged as a central concern also witnessed the emergence of the mechanical philosophy as the preferred world-view. Elsewhere I have cast doubt on the notion that the mechanical philosophy was an important factor that helped to make the new science possible. It is better read as a world view that utilized rather than spawned the new science.[8] In this section I take up the issue in the context of the foregoing history of hydrostatics.

A basic tenet of the mechanical philosophy was that the material universe consists of corpuscles of universal matter characterized solely by their size and shape and possessing a degree of motion or rest, and nothing else. This stood in contrast to the Aristotelian view according to which items in the world are what they are by virtue of the forms that animate their matter. The mechanical philosophers aimed to eliminate Aristotelian forms because they considered their ontological status to be unintelligible insofar as they were something other than matter. There are two features of the mechanical philosophy that lie behind my skepticism concerning its ability to feed into science and so be a productive force during the Scientific Revolution. First, the mechanical philosophy was based on fundamental claims about the universe in general and in its entirety whereas the sciences developed in the seventeenth century were subject-specific. Second, the mechanical philosophy was intended to characterize the ultimate make-up of the Universe. Corpuscular mechanisms were not seen as susceptible to or in need of explanation at a deeper level. By contrast, individual sciences such as mechanics, optics or astronomy involved explanations that appealed to such things as weight, light rays or planetary orbits. Insofar as such entities might well be subject to explanation at a deeper level they are what Boyle called intermediate causes, as opposed to ultimate causes of the kind sought by mechanical philosophers. Ultimate causes, unlike intermediate causes, were out of range of what could be investigated experimentally.

A large part of the story I have told in this book is a portrayal of the way in which hydrostatics emerged in the seventeenth century as an extension of the science of simple machines, which had its roots in ancient technologies and its theoretical

[8] I argue in Chalmers (1993) and Chalmers (2002) that Boyle's contributions to science did not flow from his adherence to the mechanical philosophy. In Chalmers (2009) I defend the view that the path to knowledge of atoms in modern science owed little to the philosophical tradition of atomism from Democritus to the mechanical philosophers. I distinguish intermediate causes from ultimate ones in Chalmers (2012) arguing that science is best seen as exploring the former, leaving the search for the latter to philosophers like some of the mechanical ones in the seventeenth century.

beginnings in Aristotle and Archimedes. Hydrostatics required the fashioning of the notion of pressure as a cause that needed to be added to weight. The latter cause had long been taken for granted and had come to be theorized with increasing sophistication. Viewed in this way, hydrostatics involved the articulation of intermediate causes that could be brought to bear on observable phenomena and that were accessible to experimental exploration. As such my history of hydrostatics serves somewhat as a Trojan horse as far as my position concerning the role of the mechanical philosophy is concerned. The irrelevance of the latter to the progress of hydrostatics is implicit in my story insofar as, in it, I did not need to invoke the mechanical philosophy in my characterization of those developments, except to indicate how its deployment by Descartes in hydrostatics was unproductive and how it led Boyle to distinguish between intermediate causes as they figured in experimental philosophy and ultimate causes as deployed by mechanical philosophers.

The work of Stevin fits well into the perspective I advocate. He was an engineer, skilled in mathematics but unconcerned with and uninvolved in general concerns about world-views of the kind that came to be referred to as the mechanical philosophy. His patron, Prince Maurice, had no grounds for qualms about the applicability and usefulness of the sciences developed by Stevin, including hydrostatics, because they involved intermediate rather than ultimate causes.

Pascal was a proponent of the mechanical philosophy insofar as he was sympathetic to the atomistic version of it articulated by Pierre Gassendi. For a while Pascal construed the new pneumatics and its rejection of the force of a vacuum as a decisive refutation of the Aristotelian world view and a vindication of the mechanical one. However, it soon became apparent that both the mechanical philosophers and the scholastics were divided on the question of the possibility of a vacuum, and, in any case, it transpired that the new pneumatics required only the acknowledgement that air has weight and that it is largely absent from the space above the mercury in a barometer. In any event, Pascal's development of hydrostatics in *The Equilibrium of Liquids* appealed to their weight and fluidity with no hint of the need to explain such notions at a deeper level.

There is no doubt that Boyle was a supporter of the mechanical philosophy, providing sophisticated articulations and defences of it and playing a large part in making that philosophy popular. However, as we saw in Sect. 8.3, it was Boyle who introduced the notion of intermediate as opposed to ultimate causes and used it to argue that sciences such as pneumatics and hydrostatics could be developed without a need to explain notions such as weight, spring and pressure by appeal to corpuscles possessing only shape, size and motion. The case Boyle made for his hydrostatics was independent of the ontological claims central to the mechanical philosophy. Not only is such a claim borne out by the arguments in Boyle's hydrostatic texts, but it is also a point stressed explicitly by Boyle himself.

When a mature Newton came to take a stand on the status of gravity as it figured in the *Principia* he took a position that, in effect, was in accord with Boyle's stance on intermediate causes. In the General Scholium in Book 3 Newton claimed to have 'explained the phenomena of the heavens and of our sea by the force of gravity' whilst admitting that he had 'not yet assigned a cause to gravity'. Nevertheless, he continued, 'it is enough that gravity really exists and acts according to the laws we have set forth and is sufficient to explain all the actions of the heavenly bodies and our sea' (Newton 1999, p. 943). As far as his hydrostatics is concerned we have seen that Newton assumed that liquids are characterised by their inability to withstand distorting forces. In *De gravitatione* he alluded to the fact that such a feature must arise from 'the innumerable circumstances affecting the figures, motions and contextures of the least particles' composing a liquid. However, he then abstracted from them 'since the physical cause of fluidity is not to be considered here' (Hall and Hall 1962, p. 151). When, in the *Principia,* Newton introduced the gravitation of liquids into the picture there was no suggestion that the property was in need of some deeper explanation. Insofar as the world picture that a mature Newton occasionally invoked involved particles interacting via forces acting at a distance he departed from the official 'mechanical philosophy' in a significant way. But whatever the details of that story might be, they are insignificant as far as hydrostatics is concerned. Newton's hydrostatics was unconnected with and not dependent on the articulation of a world-view. In this regard, Newton's practice was in accord with that at work in Boyle's deployment of weight and pressure in his hydrostatics and also with Boyle's explicit stand on the status of intermediate causes. It is likely that Boyle had a direct influence on Newton in this respect.[9]

In my opening chapter I anticipated that in this book I would write a history of hydrostatics and then employ it in an attempt to shed light on some key features of the Scientific Revolution. My remarks on the relation between hydrostatics and the Scientific Revolution in this section should be seen as a part of the latter project. My

[9] There are a number of factors that suggest a likely influence of Boyle on Newton in these respects. If I am right to assume that Newton became aware of the hydrostatic tradition through reading Boyle's *Hydrostatical Paradoxes* then he would have confronted an approach that took the continuity, fluidity and weight of water for granted, with no suggestion that those features were in need of a deeper explanation. Newton had certainly read Boyle's *Spring of the Air* where Boyle explicitly made the point that he was to characterise and appeal to spring while not explaining it at the corpuscular level. Boyle made the point that experimental science could and should proceed by appeal to unexplained causes such as weight and elasticity in the 'proemial essay' with which he introduced his *Certain Physiological Essays* of 1661. Boyle also entertained the idea of a scale of causes, ranging from immediately apparent causes, such as weight, through various levels of intermediate causes up to the primary causes, the 'primary affections of the smallest parts of matter'. Here is how Boyle expressed it in 'A requisite digression concerning those that would exclude the Deity from intermeddling with matter', the fourth essay of his *Usefulness of Natural Philosophy* published in 1663. 'And sure, Pyrophilus, there are diverse effects in nature, of which, though the immediate cause may be plausibly assigned, yet if we further enquire into the causes of those causes, and desist not from ascending in the Scale of Causes till we arrived at the top of it, we shall perhaps find the more catholic and primary causes of things' (Boyle 1999, Vol. 3, p. 245). This theme is echoed by Newton in an unpublished Preface to the *Principia.* 'He who investigates the laws of effects of electric with the same success and certainty [as evidenced in Newton's treatment of gravitation] will greatly promote philosophy, even if perhaps he does not know the cause of those forces. First the phenomena should be observed, then there proximate causes - and afterwards the causes of the causes – should be investigated, and finally, it will be possible to come down from the causes of the causes (established by phenomena) to their effects, by arguing a priori' (Newton 1999, p. 53). Here what Newton calls proximate causes are what Boyle referred to as intermediate, or occasionally sub-ordinate, causes.

account of the theories of hydrostatics and their relation to one another is independent of, and does not require acceptance of, my thesis about the relation between experimental science and the mechanical philosophy.

References

Boyle, R. 1999. In *The works of Robert Boyle*, ed. M. Hunter and E. Davis, vol. 14. London: Pickering and Chatto.

Chalmers, A. 1993. The lack of excellency of Boyle's mechanical philosophy. *Studies in History and Philosophy of Science.* 24: 551–556.

———. 2002. Experiment versus mechanical philosophy in the work of Robert Boyle. *Studies in History and Philosophy of Science.* 33: 187–193.

———. 2009. *The scientist's atom and the philosopher's stone: How science succeeded and philosophy failed to gain knowledge of atoms*. Dordrecht: Springer.

———. 2012. Intermediate causes and explanations: The key to understanding the Scientific Revolution. *Studies in History and Philosophy of Science.* 43: 551–562.

Cottingham, J., R. Stoothoff, D. Murdoch, and A. Kenny. 1991. *The philosophical writings of Descartes*. Vol. 3. Cambridge, MA: Cambridge University Press.

Gooding, D. 1990. *Experiment and the making of meaning: Human agency in scientific observation and experiment*. Dordrecht: Kluwer.

Hall, R.A., and M.B. Hall. 1962. *Unpublished scientific papers of Isaac Newton*. Cambridge, MA: Cambridge University Press.

Harper, W.L. 2009. Newton's methodology. In *Quantum reality, relativistic causality and the epistemic circle*, ed. W.C. Myrvold and J. Christian, 43–61. Dordrecht: Springer.

Kuhn, T. 1959. *The Copernican revolution*. New York: Random House.

Newton, I. 1999. In *The principia: Mathematical principles of natural philosophy*, ed. I.B. Cohen and A. Whitman. Berkeley: University of California Press.

Smith, G. 2002. The methodology of the *Principia*. In *The Cambridge companion to Newton*, ed. I.B. Cohen and G. Smith, 138–173. Cambridge, MA: Cambridge University Press.

Chapter 12
Hydrostatics and the Scientific Revolution

Abstract The extent to which dealings with the world as it can be observed and manipulated is dependent on the availability of concepts up to the task has not been adequately appreciated by historians of science. Detailed attention to changes that the concept of pressure underwent in the seventeenth century can serve to ameliorate that situation. The history of hydrostatics can also usefully illustrate the way in which science emerged in the seventeenth century as autonomous from mathematics, practitioners' knowledge and metaphysics. Hydrostatics employed mathematics but differed from mathematics insofar as its postulates were in need of empirical support. As far as practitioner's knowledge of hydrostatics is concerned, that did not change much during the seventeenth century. However, the level of understanding of hydrostatics was much deepened by way of theories that could explain what was taken for granted by practitioners. Finally, knowledge of hydrostatics was distinct from the accounts of the ultimate structure of the world favored by mechanical philosophers. It owed little to their attempts to construct such a metaphysics. From a modern point of view the technical sense of pressure can appear obvious. There is much to be learnt about science by recovering the way in which it became obvious.

Conceptual change of the kind represented by the emergence of the modern concept of pressure in hydrostatics was a key component of the changes that constituted the Scientific Revolution. The fashioning of the concepts of force and inertia by Isaac Newton is a development that has been well appreciated and extensively studied. But other important changes share with the change in the concept of pressure the feature of being inadequately appreciated. The changes in the concept of motion that brought Galileo and Rene Descartes to the brink of transforming pre-classical into classical mechanics is a case in point. The study of this change by Jürgen Renn and his associates at the Max Planck Institute for the History of Science has not had the influence that it warrants.[1] Another study of conceptual change stemming from the Max Planck Institute is that of the emergence of the concept of chemical compound by Ursula Klein.[2] As is the case with my account of the fashioning of the

[1] See Damerow et al. (2004).

[2] See Klein (1994).

A.F. Chalmers, *One Hundred Years of Pressure*, Archimedes 51,
DOI 10.1007/978-3-319-56529-3_12

concept of pressure, the changes described by Renn et al. and Klein do not involve deep theoretical probing behind the phenomena by invoking atoms, imponderable fluids or chemical elements and principles. Rather, they involve the construction of concepts up to the task of making possible the characterization of the world as readily accessible via observation and manipulation. Once concepts of pressure, motion and chemical compound had been fashioned, the recognition that they can so readily be applied to familiar phenomena has the effect of rendering invisible the fact that they are not self-evident givens but rather needed to be constructed.

Just as the concept of pressure was not a given, so the postulates of hydrostatics formulated with its aid were not givens. For that reason, a science of hydrostatics could not be developed as the consequences of axioms ready to hand at the beginning of the seventeenth century. Geometry and statics could plausibly be construed as sciences involving mathematical consequences of unproblematic and generally accepted postulates. We have seen how Simon Stevin attempted to construe hydrostatics in that way. The claim made by Stevin, and by Blaise Pascal after him, to the effect that water presses on a solid surface at a given depth below the surface of a body of water to a degree that is independent of the orientation of the surface, was far from evident. Both of them employed thought experiments as well as experiments that could be performed to substantiate that claim and Robert Boyle after them certainly performed a range of straightforward experiments to that end. We see here a move from the conception of a science as a body of knowledge derived from unproblematic axioms acceptable at the outset to one that recognizes the need to clarify the concepts and support the claims of a science empirically. That move, which we have documented in the case of hydrostatics, was to be a crucial element of what constituted the Scientific Revolution.

The development of hydrostatics did not end with Boyle. The advances introduced by Isaac Newton led to a new depth of understanding. Newton showed how Stevin's law and Pascal's principle follow as a consequence of what he identified as the essential property of fluids, namely, their inability to withstand a distorting force. Newton did not extend the range of hydrostatic phenomena that were subject to confirmation via observation and experiment. Rather, by identifying the essential property of liquids and showing how the range of observable and experimental phenomena follows from it, he extended and enhanced the explanatory power of hydrostatics. Insofar as that feature was typical of scientific advances of the seventeenth century, it exemplifies the move towards scientific knowledge providing understanding that was qualitatively different from, and probed deeper than, everyday knowledge and the practical knowledge of artisans.

Experiment was involved in the clarification of the concepts of hydrostatics, especially the new concept of pressure, and in supplying some of the evidence in its support. In that respect it warrants recognition as an example of the new 'experimental philosophy'. However, the comparison of pneumatics and hydrostatics has shown that there was one notable feature of experimental science that was not involved in seventeenth-century hydrostatics to a significant degree. Experiment came to be used to an increasing extent in the seventeenth century as a means of revealing phenomena not otherwise known, many of which could only be manifest in artificially contrived situations. It was sometimes the case that experimental evidence for a theory was acquired as the result of testing its predictions. Testing the claims of the new pneumatics by transporting a barometer to the top of the Puy de Dôme is a case in point. Our study of hydrostatics illustrates the point that experimental support via novel predictions was not an essential feature of that science. Hydrostatics was borne out by the extent to which it could explain a wide range of evidence confirmed by observation and experiment in an uncontrived way. This was a characteristic possessed in a particularly telling way by Newton's version of the theory which could derive the whole range of hydrostatic phenomena by appeal to the definition of a liquid. However, the range of phenomena explained by Newton was already known to Stevin before the seventeenth century began. Newton's hydrostatics was strongly confirmed empirically because of the natural, uncontrived way in which it could explain a wide range of phenomena already known. Some of the new knowledge established in the seventeenth century undoubtedly did involve novel experimental discoveries and it was sometimes the case that such discoveries came as the result of testing theoretical predictions. Our history of hydrostatics, which did not conform to this pattern, helps to put this kind of point in perspective.

It is instructive to single out three ways in which hydrostatics epitomizes the emergence of science as a distinctive form of knowledge in the course of the Scientific Revolution. Hydrostatics emerged as a science as distinct from mathematics, practitioners' knowledge and metaphysics.

There is no doubt that hydrostatics, like much of seventeenth-century science, involved mathematics in a significant way. This is particularly evident in the case of Newton's version of it. Newton was able to deploy his definition of liquids by reference to their inability to withstand distorting forces only by assuming that they possessed this property all the way down. His proofs involving spheres touching planes and other spheres at points were necessarily mathematical. However, while the formulation of theories of hydrostatics and derivations from them were mathematical, those theories differed from geometry insofar as their axioms could not be construed as unproblematic givens. This distinction was not evident in the case of sixteenth-century statics, the paradigm of scientific knowledge at the time, because its axioms could be represented as obvious and uncontentious. The representation of the principles of hydrostatics as unproblematic givens becomes untenable once it is recognized that the notion of pressure and the defining characteristic of liquids were far from obvious. The adequacy of Newton's definition of liquids was demonstrated by the way in which the known properties of liquids followed from it. Hydrostatics exemplifies the way in which the new science was empirical in a way that Euclidean geometry was not presumed to be.

We have seen that seventeenth-century hydrostatics began with Stevin's contributions to it. Stevin's insights drew heavily on his experience as a hydraulic engineer. In that respect, modern hydrostatics began with and grew out of practitioner's knowledge. However, it did not stay there. The study presented in this book has shown how seventeenth-century hydrostatics progressed in a way that deepened the understanding of the subject while adding little at the level of practitioner's knowledge. Eventually hydrostatics was to renew its links with practitioner's knowledge. In 1795 a workable version of the hydraulic press was patented by Joseph Bramah.[3] He took the modern conception of pressure for granted. It was likewise taken for granted in many of the novel technologies of the time such as steam engines and took a central place in nineteenth century physics, especially thermodynamics and fluid dynamics. The gas law, PV = nRT, presupposes the modern conception of pressure that has an intensity but not a direction. As the foregoing study has shown, the concept of pressure readily taken for granted here was simply not available until it was put in place as a result of a century of intellectual effort in the seventeenth century. Hydrostatics grew out of the practical knowledge that existed prior to the seventeenth century and eventually yielded new technologies beyond the seventeenth century. The new level of understanding that was forged in the seventeenth century was qualitatively distinct from both.

When claiming the new sciences such as hydrostatics to be distinct from metaphysics I have a specific issue in mind. It is the attempt by the mechanical philosophers to replace the Aristotelian characterization of the world in terms of matter animated by forms with one involving particles of brute matter possessing nothing other than shape, size and a degree of motion, with the elimination of forms. My point was expressed and anticipated by Boyle. The uses to which concepts such as weight, elasticity, force, and pressure were put in seventeenth-century mechanics, including hydrostatics, were quite independent of whether and how those entities could be explained at a deeper, let alone ultimate, level. They were what Boyle referred to as intermediate causes. The details of the history of hydrostatics that I have described do not involve or depend on the extent to which weight, pressure and the like can be explained in a way sanctioned by the mechanical philosophers. This feeds into my suspicion that the metaphysics endorsed by the mechanical philosophers was at a level too remote from what could productively inform experimental investigation.[4] I am inclined to see in seventeenth-century hydrostatics an example of the beginnings of the emancipation of science from metaphysics.[5]

In a modern text on the mechanics of fluids the authors, in an introductory chapter, assert that 'to say that pressure acts in any direction, or even in all directions, is meaningless' (Massey and Ward-Smith 1998, p. 13). Those authors, understandably given their purpose, take the modern notion of pressure for granted and assume that pressure is a scalar quantity possessing intensity but not a direction. However, we

[3] Bramah was born in Barnsley in the North of England as the son of a farmer. He moved to London as a young adult to make his living as an inventor, doing well for himself by devising a workable version of the flush toilet. He became a celebrated locksmith and techniques that he devised for fashioning locks led to vast improvements in the British machine tool industry. In 1814, in what proved to be the final year of his life, Bramah constructed a hydraulic press capable of uprooting trees. While supervising its first application on site he caught a chill from which he died.

[4] I develop this thesis in the references cited in Chap. 11, footnote 8.

[5] The distinction between science and metaphysics that I invoke here should not be confused with the positivist version of that distinction, according to which genuine scientific knowledge is based on secure foundations supplied by the senses. The main burden of the argument of this book is that the concept of pressure necessary for characterising liquids at the level of observation was not an unproblematic given.

have seen that the move from a common concept of pressure, from the perspective of which directed pressures are paradigmatic rather than meaningless, to the modern, technical concept of pressure, was far from straightforward. A good articulation of the notion of pressure can make it appear as an obvious given to the modern student once it is comprehended. We have seen in the course of this book how a 'good articulation' of the modern concept of pressure became possible as the result of theoretical endeavor and conceptual innovation spread over the century that separated Stevin's *Elements of Hydrostatics* and Newton's *Principia.* As Gaston Bachelard put it back in 1938, 'empirical thought is clear *in retrospect*, when the apparatus of reason has been developed'.[6]

References

Bachelard, G. 2002. *The formation of the scientific mind.* Trans. M. McAllester. Manchester: Clinamen Press.

Damerow, P., G. Freudenthal, P. Mclaughlin, and J. Renn. 2004. *Exploring the limits of pre-classical mechanics: A study of conceptual developments in early modern science*. New York: Springer.

Klein, U. 1994. Origin of the concept of chemical compound. *Science in Context.* 7: 163–204.

Massey, B., and J. Ward-Smith. 1998. *Mechanics of fluids*. Cheltenham: Stanley Thornes Publishers.

[6] See Bachelard (2002, p. 24, italics in original).

Name Index

A.F. Chalmers, *One Hundred Years of Pressure*, Archimedes 51,
DOI 10.1007/978-3-319-56529-3

Subject Index

A.F. Chalmers, *One Hundred Years of Pressure*, Archimedes 51,
DOI 10.1007/978-3-319-56529-3

MIX
Papier aus verantwortungsvollen Quellen
Paper from responsible sources
FSC® C105338

If you have any concerns about our products,
you can contact us on
ProductSafety@springernature.com

In case Publisher is established outside the EU,
the EU authorized representative is:
Springer Nature Customer Service Center GmbH
Europaplatz 3, 69115 Heidelberg, Germany

Printed by Libri Plureos GmbH
in Hamburg, Germany